Praise for Kuni

"Reading *Kuni* makes me want to dive into rural Japan. Heartbreaking in many ways, this book reminds me that leaders emerge when and where you least expect it."

—ALICE WATERS, founder of Chez Panisse restaurant, activist, and author

"Mr. Tsuyoshi Sekihara is not a scholar. But in a different era, he would have been. Readers will be surprised by the strength of words and depth of thought he produces. If society understands the value of his work, it will be a sign that it has become a little better."

—KUMI SEIKE, professor at Ritsumeikan Asia Pacific University

"We need more kuni—healing and bridging between people divided by geography. From Japan to the Americas, forgotten rural communities desperately need new paths to forge a future."

—RICARDO SALVADOR, senior scientist and director of the Food and Environment Program at the Union of Concerned Scientists

"What are the universal values common to all humanity? In a kuni—a place full of vitality and where humans have pride and free will—individuals continually engage in courageous practice and never cease to question their work. In Tsuyoshi Sekihara's new theory of kuni, we find signs of light that will help us survive in this difficult time."

—TAKAO AOKI, board chair of KODO Group/ Kitamaesen Co., Ltd.

"This is a much-needed and hugely attractive idea—or set of ideas—for overcoming the rural/urban divide, which sadly does exist and usually breeds a lack of understanding that goes both ways."

—DEBORAH MADISON, cookbook author and chef

"I am immensely proud of *Kuni*. This book is the outcome of a multi-year US-Japan rural exchange program and is a testament to how exchange programs can positively impact international dialogue and understanding and help innovative leaders see their work in new, illuminating, and fresh ways. Kudos to Richard McCarthy and Tsuyoshi Sekihara, who generously wrote this book so that others working on the rural-urban divide can benefit from their expertise and experience."

—AMBASSADOR MOTOATSU SAKURAI, president emeritus of Japan Society, Inc.

"This remarkable East meets West manifesto for the best of an enlightened globalization may bring us all back to where we should be: home. Sekihara and McCarthy urge us to return home, to defend fragile rural places that our criminal food system plunders. Best read over rice dishes."

—CARLO PETRINI, founder of Slow Food

"In *Kuni*, we witness how valuable community-level exchanges between the United States and Japan can be to facilitate sharing of lessons learned and best practices on common challenges facing our two countries. As two innovative leaders of their respective societies, McCarthy and Sekihara not only help shed light on a key issue of concern shared by the United States and Japan, but also encourage us all that important ideas and experiences are being shared between the two countries to strengthen urban-rural ties and help rural communities thrive. Their collaboration is an exceptional manifestation of sub-national exchanges that form a strong foundation for an enduring US-Japan relationship, and will also be a significant resource for rural revitalization efforts in communities around the world."

—KAZUYO KATO, executive director of Japan Center for International Exchange (JCIE/USA)

"Richard McCarthy is one the most playful and ground-breaking thought leaders in the Americas."

—GARY NABHAN, ethnobotanist and author

KUNI

KUNI

A Japanese Vision and Practice for Urban-Rural Reconnection

TSUYOSHI SEKIHARA and RICHARD McCARTHY

Foreword by Kathleen Finlay

North Atlantic Books

Huichin, unceded Ohlone land
aka Berkeley, California

Published by
North Atlantic Books
Huichin, unceded Ohlone land
aka Berkeley, California

Cover photo © skywings00 via Shutterstock
Cover design by Susan Zucker
Book design by Maureen Forys, Happenstance Type-O-Rama

Printed in the United States of America

Kuni: A Japanese Vision and Practice for Urban-Rural Reconnection is sponsored and published by North Atlantic Books, an educational nonprofit based in the unceded Ohlone land Huichin (*aka* Berkeley, CA) that collaborates with partners to develop cross-cultural perspectives; nurture holistic views of art, science, the humanities, and healing; and seed personal and global transformation by publishing work on the relationship of body, spirit, and nature.

North Atlantic Books' publications are distributed to the US trade and internationally by Penguin Random House Publisher Services. For further information, visit our website at www.northatlanticbooks.com.

Library of Congress Cataloging-in-Publication Data

Names: Sekihara, Tsuyoshi, author. | McCarthy, Richard (Activist), author. | Finlay, Kathleen (Activist), writer of foreword.
Title: Kuni : a Japanese vision and practice for urban-rural reconnection / Tsuyoshi Sekihara and Richard McCarthy ; foreword by Kathleen Finlay.
Other titles: Japanese vision and practice for urban-rural reconnection
Description: Berkeley, California : North Atlantic Books, [2022] | Includes bibliographical references and index. | Summary: "Tsuyoshi Sekihara and Richard McCarthy both have a deep respect for the environment and have immersed themselves in their communities to better educate others on how to create a more sustainable lifestyle"— Provided by publisher.
Identifiers: LCCN 2022011379 (print) | LCCN 2022011380 (ebook) | ISBN 9781623177317 (paperback) | ISBN 9781623177324 (ebook)
Subjects: LCSH: Cooperative societies. | Rural-urban divide. | Rural-urban relations.
Classification: LCC HT381 .S47 2022 (print) | LCC HT381 (ebook) | DDC 307.74—dc23/eng/20220601
LC record available at https://lccn.loc.gov/2022011379
LC ebook record available at https://lccn.loc.gov/2022011380.

1 2 3 4 5 6 7 8 9 KPC 26 25 24 23 22

This book includes recycled material and material from well-managed forests. North Atlantic Books is committed to the protection of our environment. We print on recycled paper whenever possible and partner with printers who strive to use environmentally responsible practices.

Dedicated to friend and author
Ryoko Sato (1962–2019)

Contents

Foreword

The moment Sekihara-san entered my house, I knew I was in for an intellectual and spiritual journey. I was not disappointed.

Invited by my friend and colleague Richard McCarthy, I hosted Sekihara-san to hear and share his thoughts about the connections between rural and urban places with a handful of thought leaders working on increasing the ties between those often disparate worlds in New York State. I expected an evening of learning, inspiration, and cultural exchange. But being in Sekihara-san's presence is something more. Even through translation, one immediately recognizes that his is no ordinary soul. He gives off the wise elder aspect that one might expect, but he also exudes a sort of chummy feel—looks you right in the eye and talks very directly about challenges and hopes as if you've known each other for years. The cumulative effect is that everything he is saying just *makes so much sense.*

I've spent the last decade running the Glynwood Center for Regional Food and Farming with a mission to create a regional identity for the Hudson Valley that centers on farming. Our hypothesis is that if we recognize the central role that farming has and still plays in this region—for our health, for our economy, for our environment, and culturally for our communities—then the sector, threatened by industrialized food and development pressure, has a chance to thrive. We use food and farming to build local economies, to nourish our communities, to attract and entertain our visitors.

This mission to highlight and promote farming and local agriculture requires two parallel strategies: coalescing our local communities and strengthening its ties to that amazing, ever-changing urban mecca that is New York City. Sekihara-san affectionately calls the metropolis "The Beast." His use of language here is telling. It's provocative, for sure, but also pictorial and imaginative. Sekihara-san describes the city as a big, hungry living organism that is often hard on its residents and, he posits, strengthening ties to its more serene cousins in the countryside will serve as a soothing and calming salve.

He's not wrong. Facilitating that connection requires a complex approach, one that is laid out beautifully in this book. Glynwood's strategy has been similar.

We train new-entry farmers in regenerative agriculture. This new generation of land stewards is not your grandfather's farmers—in fact, most of them that we have trained are women. They are driven by social values of community, justice, and sovereignty. They have deep respect for the land. They are often college-educated and do not come from farm families—quite the opposite. Many come from New York City or other cities. They are the builders of *kuni*—learning to farm, but much more than that, they want to foster communities and form deep, trusting relationships. They are using food and farming as a cultural tool: one that brings its residents together; one that helps to feed their neighbors in need; one that expresses the environmental values of regeneration; one that helps bring people agency over what they eat beyond what is available in big-box stores.

We also do a lot of experiential teaching at Glynwood, hosting a range of experts to help diverse audiences, from CEOs to schoolteachers, about why supporting a regional food system is so important. And we build coalitions. This work is complicated, unpredictable, and time-consuming. But by fostering small networks of people with very similar ambitions, we create a web of interconnectedness that can stand even pretty big shocks, like the COVID-19 pandemic. Community-supported agriculture may be one of the most tangible models in the United States of the types of convents between humans that you will read about in this book. A pact—I will grow food for you, you will pay me—that turns into so much more: I will nourish you with taste; you will savor it. And through that act of deliciousness, we are connected.

All of this is working; we are indeed saving farms, getting young people to leave "The Beast" and work and live and play in rural settings, building relationships that form healthy communities. We are building kuni. But we could be doing more—and in the following pages there are ideas that *just make so much sense.*

My hope is that this book sparks the imagination of those who have not had the pleasure of sitting down with Sekihara-san to take seriously these calls to action, help identify the change-makers that are ripe for carrying them out, and influence the policy-makers and philanthropists that can resource those efforts.

We need efforts like these to become a more just, wiser society. Recently, I learned about an extraordinary effort in Zimbabwe called Friendship Benches. In public parks, trained elders hang out on discreetly marked benches that invite community members in need to sit and receive a mental health check-in. It's a brilliant idea, but I couldn't help thinking that if we manage to get some of these things right, if we build kuni, wouldn't this happen on its own? Wouldn't we seek out elders for advice and welcome an exchange? Wouldn't there be friendship benches that don't need any markings?

Hosting Sekihara-san was like sitting down on a friendship bench with him. I am better for it; the work at Glynwood is better for it. And thanks to Richard's superb and passionate act of ambassadorship in cowriting this book, you will be the better for it too.

—KATHLEEN FINLAY

Glynwood Center for Regional Food and Farming

Acknowledgments

We are profoundly grateful to the Japan Foundation Center for Global Partnership, Mitsubishi Corporation (Americas), R&R Consulting, ANA Holdings, and United Airlines for their support for the international encounters that have made this book possible.

We are especially grateful to:

The Japan Society (and specifically **Betty Borden,** Director, Innovators Network, and **Fumiko Miyamoto,** Senior Program Officer, Innovators Network, for their creative partnership and extraordinary translation skills) to weave the threads of thought between Japan and the United States;

The Japan NPO Center (and specifically **Katsuji Imata,** member, Board of Directors; **Shinji Nagase**, staff member; **Kazuho Tsuchiya,** Senior Leader) for their visionary leadership to design and facilitate international learning exchanges with the Japan Society;

Atsuhisa Emori, General Manager, Nippon Taberu Journal League in Hanamaki, Iwate Prefecture; **Kenji Hayashi,** Co-President, FoundingBase in Tsuwano, Shimane Prefecture; **Connie Reimers-Hild,** former Chief Futurist, Rural Futures Institute, University of Nebraska in Lincoln, Nebraska; **Johnathan Hladik,** Director, Policy Programs, Center for Rural Affairs in Lyons, Nebraska; **Savanna Lyons,** Principal, May Day Consulting and Design in Huntington, West Virginia; **Ryoko Sato,** Associate Professor, Ehime University in Matsuyama, Ehime; **Taylor Stuckert,** AICP, Executive Director, Clinton County

Regional Planning Commission and cofounder, Energize Clinton County in Wilmington, Ohio; **Junichi Tamura,** Chief Director, Next Commons Lab and Director, Tono Brewing Company in Tono, Iwate Prefecture, for helping us learn, share, and grow together;

Kathleen Finlay, President, Glynwood Center for Regional Food and Farming in Cold Spring, New York; **William Morrish,** Professor of Urban Ecologies, New School in New York City; **Janet Topolsky,** former Executive Director, Community Strategies Group, Aspen Institute in Washington, D.C., for helping us to place kuni in the spectrum between theory and practice in the United States; and **Bonnie Goldblum,** for capturing the exchanges on-site in Japan, and for helping to shape the concepts for North America.

Kuni

Definition *noun*

1. A nation
2. An ancient community that is small, but independent

KUNI MANIFESTO *By Tsuyoshi Sekihara*

A *kuni* can be created anywhere through the determination of one person. Even a hamlet on the verge of extinction can revive its future through the will of one person.

As we despair about shrinking rural areas and are intimidated by cities growing to a monstrous scale, where are we supposed to go? We are dazzled by image, enslaved by fads, and compelled to buy, buy, buy.

Can we stop buying and start making? Stop viewing and start finding? Stop speaking and start listening? When did these things become so difficult?

Kuni is the name for a new community

A *kuni* has at least 500 people but no more than 2,000. What is too small? Small rural communities in rapid decline lose diversity and experience the rise of mutual surveillance and small dictatorships. What is too large? In gigantic cities that grow exponentially, people become part of the scenery.

Everyone is equal in a *kuni*

In a *kuni*, people share in the abundance and the hardships provided by the land. This brings people together and forms a sense of "we." There is no discrimination based on race, religion, income, or ideology.

Kuni is equipped with a Regional Management Organization (RMO)

The RMO is a democratic organization that takes care of small public services. These services include diverse programs such as education programs to enhance children's five senses, programs to maintain the health of the elderly, and administrative tasks. The RMO also serves as an intermediary for repeat visitors, nurturing regional economies using local resources, implementing small-scale transportation services, and conserving nature. By providing these services, young staff at an RMO also hone their skills in comprehensive policymaking.

Kuni is a link between residents and repeat visitors

A *kuni* is not solely made up of residents, but becomes a community when it includes repeat visitors from cities, who feel affection and a sense of belonging to the land.

In a *kuni*, people meet and talk about the abundant harvest, the damage done by a typhoon, a newly born child, a death in the community, the taste of vegetables picked that morning, the taste of spring water from the mountain, the scent of the wind, and the cold rain from yesterday.

Do you think a *kuni* like this is a dream? All you need in the beginning is a small will.

Kuni is self-sufficient

Kuni requires an economy based on self-sufficiency and serves as a place of mutual benefit. For repeat visitors from cities, it serves as a kind of insurance policy against ongoing urban stress and anxiety. For residents, repeat visitors bring economic activity and new relationships.

Life in a *kuni* is circular and forms a beautiful spiral

The basis of *kuni* is circularity, where production and consumption are in balance. A life with circularity forms a beautiful spiral over time.

Kuni embraces the whole person

In a *kuni*, people are not cogs in a machine.

The time for *kuni* is now

Cities are growing bigger, and wild and savage, and rural places are disappearing.

Kuni is a place for young people who seek interconnectedness

Local governments are downsizing and are no longer capable of providing public services. An RMO can provide these services, and through work in an RMO, young people can help shape the future.

Be the Sisyphus of our time. Rise up.
This is the beginning of a quiet revolution.

クニ

定義　　1. 国家
名詞　　2. 古代の共同体、小規模だが独立している。

クニ・マニフェスト

文：関原剛

クニは、最初のひとりの意志さえあれば、どこにでも作り出せる。いまは滅びゆくように見える集落でさえ、ひとつの意志から始まって再び呼吸をとりもどす未来がある。

縮む地方に絶望し、怪物のように大きくなった都市に怯えるだけなら、わたしたちは一体どこへ行けばいいのだろう。虚像に踊らされ、流行の下僕になり、わたしたちは「買いつづける」しかないのだろうか。

「買うだけ」ではなく「つくる」こと、何かを「見せられる」だけではなく「見つけること」、一方的な送信ではなく、人が人の前に立って話すこと、そんな簡単なことが、いつからこれほど難しくなったのだろう。

クニとは新たな共同体の名前である

クニは500人から2000人以下の人々で構成される。小さすぎる農村の共同体は急激な衰退によって多様性を失い、相互監視と小さな独裁が現れる。大きすぎる場合は、巨大都市が過剰肥大を続け、人々は背景化してしまう。

クニは皆が平等である

クニでは土地がもたらす豊饒と試練を共有する。これが人々を引き寄せ、「われら」という感覚を作り上げる。人種、宗教、貧富、イデオロギーによる区別や差別は一切存在しない。

クニは地域運営組織(RMO)を持つ

RMOは、小さな公を担う民主的な組織である。小さな公には、子どもたちへの五感教育、高齢者の健康維持、様々な地域活動のための事務処理等が含まれる。また往還者とのつなぎ、地域資源産業の育成、小さな公共交通、環境の保全も行う。小さな公の実施によって、地域運営組織（RMO）で働く若者の総合的な政策能力が向上する。

クニでは定住者と往還者が繋がる

クニは定住者だけでつくられるものではない。土地に愛情と帰属感をもって都市から往還する人々との共同によって初めて現れる共同体である。

クニの人々は会って話す。みのりの多さ、台風の被害、子供が生まれた話、仲間が亡くなったこと。今朝の野菜の味、山で飲んだ泉の味、風の匂い、昨日の雨の冷たさを話し合う。

このようなクニは夢だろうか。いいや違う。それは具体としての未来である。はじめに小さな意思があればよい

クニは自給自足である

クニの経済は自給自足で、互恵的である。クニの自給力は、都市生活のストレスと不安に 対応する保険的な力を持つ。往還者はクニに経済活動と新たな関係性をもたらす。

クニでの生き方は円環的で美しい螺旋を描く

クニの基本は「円環」であり、生産と消費の均衡がとれている。
円環的な生活を送る人は、時間の経過とともに、美しい螺旋を描く。

クニは個々の人間を大切にする

クニでは、人間を部品化しない。

クニは現代だからこそ求められている

都市は膨張を続け、野生化し、地方は消滅している。

クニは他者とともに生きたいと願う若者たちのための場である

縮小しつつある地域行政はもはや細やかな公を行う力が無い。
RMOが代わりにサービスを提供し、若者が未来を築く手助けをする。

現代のシーシュポスを目指せ。 今こそ立ち上がれ。
これは静かな革命の始まりである。

When residents left hurriedly for urban opportunities in the postwar period, they often abandoned many keepsakes, including photographs like this one of life before the great decay. Many have been retrieved, saved, and placed in the community museum's archives.

The natural beauty of rural Japan's mountainous rice villages continues to be one of the primary assets on which locals can still trade. Located in the mountains, the community is only twenty minutes away from the Sea of Japan.

One of the commitments the Regional Management Organization makes to the elderly population of the region is to deliver food during the winter, when it is unsafe for seniors to venture out for food.

The Regional Management Organization trains schoolchildren and young volunteers to become knowledgeable and proud of the region's natural beauty and biodiversity. They conduct forestry tours as part of the agritourism programming.

In Japan, rural communities possess beautiful yet dilapidated housing. Many are one- and two-hundred-year-old farmhouses. In rural Joetsu, they are slowly being renovated to serve current community needs: as community centers, short-term rentals, restaurants, museums, and housing for those who remain.

This once-decaying farmhouse has been transformed into a high-design café and restaurant to feed visitors and to feature the wood and artisanal woodworking skills in the community.

The annual crow hunting festival is indicative of the traditional community rituals that have been revived, together with traditional agriculture. Repeat visitors gain access to festivals like this one as part of the Rice Covenant.

The Regional Management Organization's Rice Covenant provides repeat visitors from the cities with a relationship deeper than that afforded a general consumer. Pictured here, weekend visitors gather and celebrate with local farmers after contributing their volunteer labor for the rice harvest. They return for numerous activities important to rural life: the dredging of irrigation canals, thinning of the forest, and planting.

An Exchange Economy That Works as Insurance

Anxiety-based Industry

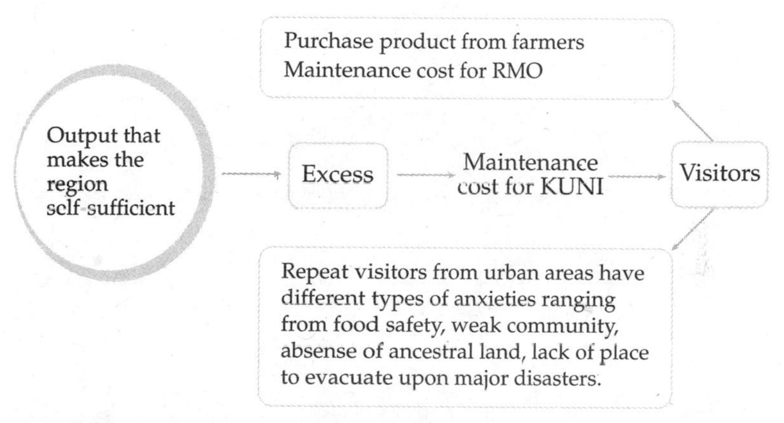

This chart describes how and what motivates trade with the outside world for both the local farmers and the repeat visitors. Not only does the drawing describe how the excess production provides earned income for the administration of the community (via the Regional Management Organization), but also that it comes only after enough is produced for local consumption.

Umeboshi plums are cured using traditional methods, with no chemical additives. They are sun-dried and packed by hand. With quality superior to what is otherwise available via industrial channels, the Regional Management Organization organizes both superior quality and opportunities for repeat visitors to be part of the process. This provides ample opportunities to trade on social and natural capital.

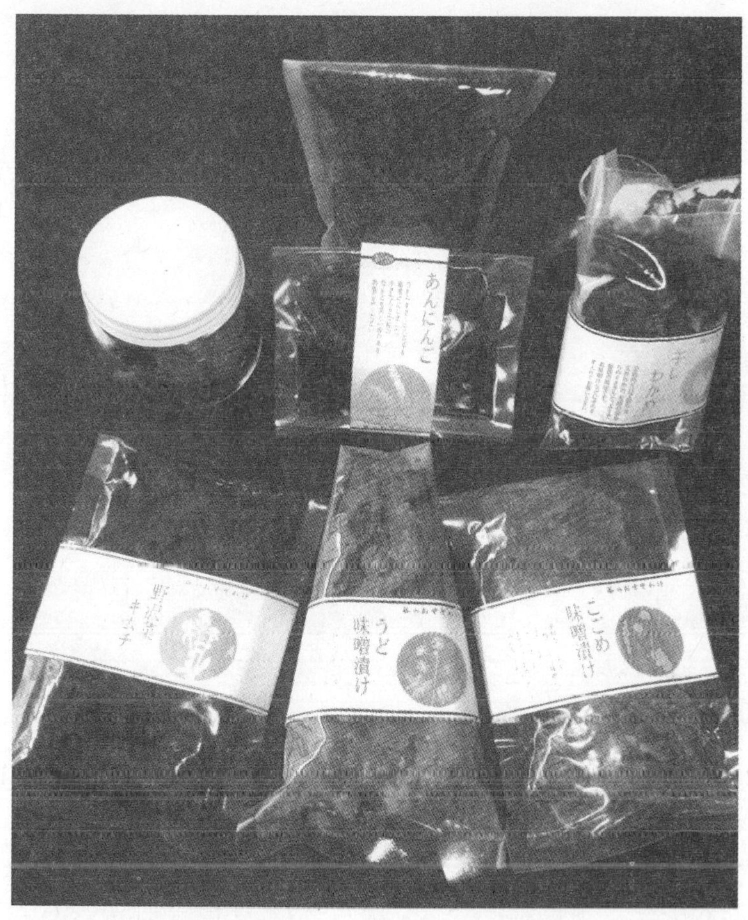

These are some of the pickled products that the Regional Management Organization's commercial operations offer to consumers. In keeping with high Japanese standards for packaging and presentation, the products pictured are no exception.

1

AN OUTSIDER ENCOUNTERS KUNI

Richard McCarthy

In the Village

The moment my feet touched the ground in Tsuyoshi Sekihara's village of Nakanomata, I was struck by the proportionality of the place. Very quickly, I understood his passion for right-sized communities. You can walk the circumference of the village within an hour. Wooden farmhouses, hundreds of years old, cluster beneath the canopy of trees along the river that runs from the mountains to the Sea of Japan. The river and its thousand-year-old manicured tributaries provide irrigation for terraced rice fields that define the rhythm of work, celebration, and nourishment in the region. One of my Japanese traveling companions gushed: "I'm not sure if I've ever walked through a functioning traditional rice village as intact as this one. I've only read about them."

Sensing that we were getting swept up by the romance of village life, Sekihara quickly reminded us: "The severe shortage of youth to work the fields, trim the forests, and produce a large enough future population means that there is little margin for error." As cited in chapter 5, media accounts relentlessly report that this is the story throughout Japan. Demographics do not favor a future for enchanting small places like Nakanomata.

Even if the stars align (as they have in this beautiful mountainous seaside corner of Japan), survival will be difficult. The building stock is breathtaking: wooden, weathered, yet perfect. Some houses are renovated while others are in desperate need of attention. With $10,000, you can become a proud owner of an authentic Japanese farmhouse. Elsewhere in Japan, they are available too. So, what is different here? Leadership and a coherent set of operating principles match Sekihara's wild ideas with practical solutions.

My Search, My Disaster

My search for kuni began long before I even met Tsuyoshi Sekihara or had heard the term *kuni.* In 2005, while reeling from the cumulative effects of the Hurricane Katrina disaster on my community, work, and family, I was searching for pathways out of the mess in which we found ourselves. We spent four months in exile in Houston. There we were: three generations hunkered down in a townhouse, strategizing the return home to New Orleans. Most waking hours were fixated on the disruptive trauma that the storm inflicted on New Orleans and the surrounding region. Would it be a one-off event or a prescient glimpse of much deeper structural crises to come?

My concern grew to obsessive proportions. Are we no longer capable of solving problems? Jared Diamond's book *Collapse: How Societies Choose to Fail or Succeed* had just been published. However, my reading did not stop there. With normal sleep nearly impossible, I was plowing through older titles, mostly science fiction: Octavia Butler's *Parable of the Sower* (1993) and George R. Stewart's *Earth Abides* (1949). These books portray societies that are in the process of no longer functioning. In *Earth Abides,* due to a pandemic, most people are simply gone.

It is this fundamental and hopeful contribution to problem solving that I immediately recognized in Sekihara's work: If the prevailing paradigm provides answers to all questions with gestures that are large-scale, then what is to become of forgotten places too small to be deemed necessary to preserve? These questions seem more urgent than ever in the wake of the rippling effects from the COVID-19 pandemic.

Thinking back to August 2005 in New Orleans, I remember that everyone was rushing around to batten down the hatches in advance of landfall. Stopped at a red light en route to the hardware store, I sat there in my car. There were no other cars around. I thought to myself, "When is it acceptable to ignore societal rules and norms, like traffic lights? Now? When do the social obligations we have for one another no longer matter? Does the social contract simply dissipate?"

After the storm, desperate images were broadcast all over the planet. Few words were necessary to convey the prevailing message: If this scale of abandonment can happen in North America, then who is safe? Many critical problems facing New Orleans predate the hurricane: poverty, systemic racism, chronic diseases, violence, and so on. However, the scale of the disaster raises key questions that we are still grappling with: For one, do US decision-makers take the vulnerabilities that come with climate chaos seriously? Second, in choosing to rebuild after disaster, on what model should recovery efforts be based? It is one thing to preserve and rebuild as it was once before. No one wants to give up ground, to turn one's neighborhood over to a designated flood plain. And yet, do we learn from disaster? Has the ethos of my home, my castle reached its limits? Do communitarian impulses need to be explored, and if so, who will pay for those who sacrifice their property and livelihoods for the greater good? These remain, largely, unanswered questions.

I witnessed so much good in the subsequent weeks and months too. Mutual aid was rampant. My vantage point was with the individuals and families involved with the farmers markets I had helped establish a decade earlier. They had taken the risks to reach out and build figurative bridges to others long before the storm. As a result, they began the disruptive period following Hurricane Katrina ahead of the curve. Despite the trauma, they did more to help one another, not less. These farmers, fishers, shoppers, and chefs had become a community of interest. Strangely, it was as if the years leading up were dry runs before the flood.

This is the pattern I was trying to understand and to describe: The nation was universally horrified by the suffering, the incompetence, and the fragility of ordinary people's lives in America. However, it was impossible to assemble a consensus that it is everyone's responsibility to right these wrongs. This is what I mean by the social contract in disarray. Meanwhile, on the ground, the community of urban shoppers and rural farmers had learned how to care for one another, at some level, through weekly interaction at farmers markets. This is the community built upon the pathway between urban and rural spaces.

The more I learned about Sekihara's praxis, the more I came to believe it is a gift to the wider field of community development and, quite frankly, to anyone who is searching for a sense of place. The parallels between his work and mine became clear once I visited his village. What echoed in my head, as he spoke, is a passage from the nearly forgotten 1957 tome *The Breakdown of Nations* by Leopold Kohr:

> In miniature, problems lose both their terror and their significance, which is all that society can ever hope for. Our choice therefore seems not between crime and virtue but between

> big crime and small crime; not between war and peace, but between great wars and little wars, between indivisible total and divisible local wars.[1]

Kohr reads as a modest proposal. His sentiments seem worlds away from the utopian works he apparently influenced, for example that of his colleague E. F. Schumacher's *Small Is Beautiful.* Speaking from personal experience, I did not feel particularly utopian after Hurricane Katrina. If anything, I felt dystopian. However, there is something hopeful and logical in Kohr's remarks that dovetails so closely with my experience of balancing urban with rural interests, farmers with consumers. Similarly, Sekihara searches for that right-size community in which problems can be solved with greater ease precisely because of the manageable scale.

Road Signs

Before I step aside for Sekihara to describe his journey and his wild ideas, here are some of the names and acronyms that feature prominently in the book.

- **NPO:** In Japan, this is the acronym for nonprofit organizations. In this regard, Japan is much like the United States, where we name these legal entities in the civic sector. For the rest of the world, these tend to be called NGOs, for nongovernmental organizations.
- **RMO:** This stands for Regional Management Organization. Sekihara describes his community's NPO as an RMO. This is the organizational structure that delivers social services previously provided by local government and also launches new initiatives to harness the creativity it will take to forge a regional future. While there are

other RMOs in Japan, few take regional management to such depths as Sekihara's. As Sekihara describes in chapter 7, in order to be successful, an RMO must address twelve key functions:

1. Maintain the well-being of residents
2. Maintain, preserve, and pass down traditional folklore
3. Keep aging populations healthy
4. Provide small-scale public transportation
5. Educate schoolchildren about the region
6. Protect nature (preserve farmland and forests)
7. Create local industry using local resources
8. Implement projects commissioned by the local government
9. Engage urbanites and organize opportunities to visit
10. Serve as an intermediary
11. Provide comprehensive administrative functions
12. Nurture human resources

- **Joetsu City:** This is the city of 200,000 residents located in Niigata Prefecture, three hours from Tokyo and overlooking the Sea of Japan. Aside from very occasional nautical skirmishes with North Korean ships, this is a calm corner of Japan: no tsunamis, few earthquakes, but plenty of snow in the winter.
- **Kamiechigo:** In an earlier era, this is what the Japanese called Joetsu. No one uses this name anymore, except for Sekihara, to name the mountainous region west of Joetsu's urban corridor. This is where he lives and works. With so many of the small villages merged with one

another's administrative offices, this rural edge of Joetsu City had lost its sense of place. By conjuring up the past, Sekihara revives a traditional rice region's sense of itself.

- **Kamiechigo Yamazato Fan Club:** This is the name of the Regional Management Organization Sekihara founded. Though its name honors the past, the RMO eyes the future by building a local and visiting fan base for the region. Working regionally, the Fan Club seeks to overcome competition between villages hungry for resources. Fan clubs are a common organizing model in Japan. Often associated with a mascot, corporations, towns, and even ill-defined regions can have a fan club.
- **Repeat visitors:** Whereas many Japanese leaders are beginning to recognize the potential of rural tourism, Sekihara seeks lasting relationships with visitors. These are the repeat visitors who become investors in a place, rather than simply consumers of disruptive mass tourism.
- **J-turn:** This demographic concept describes an individual who leaves home for elsewhere (usually a bigger city) and who returns to a place like home, but not home. Whereas a U-turn describes a return home, the J-turn differs in that there is no return to the established patterns of home. Moreover, the individual's familiarity with the patterns of this new location, yet without the personal history, places the individual at a strategic advantage of understanding how to navigate the space. This concept shapes public discourse in Japan.
- **The Squad:** This is the shorthand for the national program that places corps of young people in rural places to learn, earn a modest stipend, and repopulate declining regions with youthful energy.

- **Autonomy:** Sekihara speaks of this concept frequently; however, it is important to remember that the autonomy he craves is from the grips of local government—village mayors whose imagination and relevance faded years ago, yet whose positions remain.
- **Kuni:** The central principle of this book, this is the goal that RMOs are meant to deliver. As discussed elsewhere in the book, it is the ancient term for nation (or more specifically, the Japanese nation). Sekihara has appropriated it to mean "community." Importantly, this community is open to outsiders who lean in to contribute as repeat visitors.
- **Right-sized community:** Size determines how communities function. If too small, life can be stifling. If too large, it becomes anonymous. Sekihara strikes for the right size in order to balance freedom and belonging. This question of scale is largely dismissed as irrelevant. Finally, someone takes the question of scale head-on, rather than simply accepting the ethos that bigger is always better.
- **Nakanomata:** This tiny village on the outskirts of Joetsu is home to Sekihara's operations. It is one among approximately twenty-five villages that are part of his network.

The Loneliness of Working Together

Nakanomata is the home base for Sekihara's operation. The Fan Club works out of a collection of small public sector buildings, long since turned over to the organization. On the day we arrived in October 2018, we began our tour at the headquarters. By we, I mean the Japan Society Innovators' Network delegation. I was fortunate to be able to learn and travel with a remarkable group of American rural leaders who had hosted Sekihara during his previous year's tour of North America. Also searching for new models to reposition rural America, they (like me) were in Japan to learn.

Whereas social enterprises hinge on the ability of an innovation or technique to provide a shortcut to addressing critical societal issues, Sekihara's story caught my attention specifically because he describes that there are no shortcuts. When we discussed this matter on-site, Sekihara insisted, "The work is slow and hard, like Sisyphus." This made me smile. Sekihara was channeling one of my favorite food slogans from the New Orleans snowball stand Hansens Sno-Bliz: "There are no shortcuts to quality."[2]

Sekihara then went on to describe his strategy to build webs of trust among locals and visitors through action. These matter more than words. From what I knew from my own work to rekindle urban-rural linkages, I nodded in agreement.

When pushed to consider the practical relevance of his work outside Japan, Sekihara acknowledged that he wonders whether Japanese audiences (who seem stuck on gigantism) embrace his ideas. With remarkable candor, he then went on to describe the isolation he has felt for two decades. Yes, he has landed important local contracts from the municipal administration of Joetsu; and yes, he assembled a team of local leaders, whom he first advised when they were young interns and students, who would later run for local office. These two decades in the wilderness do not sound particularly isolating, do they?

This led us to a subject discussed far too rarely among civil society professionals: the loneliness of executive directors. Here too we found that we stand on common ground. To be successful, executive directors must generate excitement and an air of certainty, if not inevitability (even if the future is anything but certain). Daytime confidence is counterbalanced with nighttime bouts of uncertainty—everything from anxieties about the soundness of a project to doubts for genuine community support and sources for next year's budget.

This is the founder's gamble. Grow your organization quickly enough to attract supporters and collaborators. An original approach to solving social problems or personal charisma can launch an effort. However, charismatic leadership works only for so long. If the founder dominates the show for too long, charisma can repel promising leaders who determine that the founder leaves too little room for personal and professional growth. It is no small feat to balance the appetite of emerging leaders, the ego of the founder (for which one is always present), and the financial business model to sustain the organization. Many organizations experience difficulty transitioning to the next stage of development, in which leadership and systems are routinized. The same can be said of political organizations, governments, and corporations.

For the NPO, transparency helps those who fund the organization and its beneficiaries hold the organization accountable. This is where governance bodies, like boards and committees, come in. At the same time, NPOs also demand some degree of opacity in order to get the work done with discretion and to preserve the magic of charisma.

For instance, if the vulnerable seniors who come to depend on Sekihara's services doubt the viability of his organization, it may interfere with his ability to win their confidence and collaborate with his programs. This is not to suggest that NPOs should purposefully obfuscate. However, it may be necessary to shield those whose support you need from the uncertainty. Sekihara described how this balancing act (of transparency with opacity) tormented him. He leads a community into unknown terrain, forging partnerships, inviting in outsiders, and yet, somehow, he hides the risks.

In these conversations, I came to admire his discipline and modesty. He took leaps of faith two decades earlier to breathe life into a region abandoned by the state and the economy. The NPO rose to deliver services once provided by the local government—to feed the elderly, promote tourism, develop the economy, train youth, manage the forests, and more. For those engaged in this type of work, this act of carving out a lonely corner filled with doubt and concerns is par for the course. I know this corner all too well. For me, it was this conversation that truly cemented our bond.

Throughout this book, Sekihara describes his personal journey and the evolution of his ideas in his chapters. As someone working in North America, I feel strongly that we too often lock ourselves into silos that separate economic development from social services, and urban from rural. What also strikes me as important is this: While Sekihara experiments in ways we may recognize as indicative of social enterprise, his intentions are not to privatize formerly public services. Rather, he models the RMO on, and as a replacement organization for, local government.

Kuni = Community

The purpose of the RMO is to create the preconditions to reach kuni. What is kuni? Historically, it is the Japanese character for "the nation." This stems from the name of the palace—Kuni-kyō—that served as the capital during a brief yet important period in the eighth century CE. This was a critical turning point for Japan. Warring nations became one, centered around the palace in what is today Kizugawa in Kyoto Prefecture. While kuni was abandoned for other locations, it has come to symbolize a period of important cultural and political growth for Japanese identity. It is cherished as something of a golden age.

But Japanese characters can be interpreted differently. For instance, while *kuni* (国) means "nation," it can also mean "the national community," or simply "community." Sekihara imaginatively appropriates *kuni* to mean "community," giving it contemporary relevance. This is in keeping with other successful social movements, which often utilize symbols of the distant past to craft creative ideas for the future, especially if the present is devoid of direction and purpose. Consider the importance of myths and symbols in anticolonial movements, like the Native American Ghost Dances in the late nineteenth century and the Gaelic revival in early-twentieth-century Ireland. In each, participation in group activities—like dances, poetry, and plays that pay homage to the past—brings the community closer by purging the present of alien ideas and corrupting practices.

I pressed Sekihara on this point. I asked, "Is the RMO just another gatekeeper?" Having encountered gatekeepers in public housing developments in New Orleans, I became concerned. Though not necessarily a bad thing, gatekeepers accrue power by negotiating the terms with which outsiders, and especially outsiders with financial resources, engage with locals. Problems occur when they fail to share this power, make decisions for others without including them, or keep good ideas out.

I attempted to explain this tension. Just when I began to think I had not been able to communicate effectively, Sekihara led me to the historic entrance to the village. In places of transition, the Japanese erect stone figures called *Sae no kami* or *Dōsojin* (earth gods). These are meant to protect travelers on a path or on bridges from epidemics and evil spirits. Placed at the entrance to a village, another place of transition, they protect the village from harmful effects from outside. These too had been cleaned up and cared for by the RMO; however, the point he was making is this: The RMO is like a Sae no kami. It keeps out harmful influences.

At this juncture, I began to wonder if Sekihara's kuni actually is an experiment in nativism that eschews all contact with the outside world. We walked to the renovated Shinto shrine, one of the RMO's major accomplishments. The Fan Club had taken on the work to preserve this beautiful structure that had, like nearly everything else in the vicinity, fallen into disrepair. From the shrine, we walked to the village's original (or symbolic) boundary. He pointed for emphasis. The RMO should work like these symbolic markers: to keep out harmful ideas and invite in helpful ones. Just when I was poised to dismiss the entire project, he saved it with the words: "Invite in helpful ideas." I asked, what sort of helpful ideas?

To this, he suggested that I look at the booklet that he had handed us all earlier in the day. It includes a Fan Club cultural calendar featuring examples of world-renowned musicians and performers, from Tokyo and elsewhere, which the RMO had somehow lured to Nakanomata to perform. Recognizing that he had made his point, he smiled. Sekihara then went on to describe how rarely the local population ever gets to enjoy the kinds of profound cultural activities that had lured him to Tokyo as a younger man.

I left the village assured that kuni seeks ties with the helpful influences from the outside world by trading on and reinventing the best elements of an isolated rural place's past. This balance between bonding and bridging is what makes kuni so captivating. After all, it is this very tension between the two that had brought me to community organizing through food, between town and country, at approximately the same time he had begun his journey in rural Joetsu.

An Imagined Community

Sekihara appropriates the idea—an imagined community—that has become important to the study of nationalism. He takes this idea and then uses it to understand the ramifications of scale. This should strike a chord for those drawn to the promises of bioregionalism and one of its practical manifestations: the reinvention of regional food systems. If the political project of the eighth-century kuni enabled arts and civic life to flourish within the context of rigid and hierarchical authority, it did so around an imagined national community—that of the islands that make up Japan, and not just of one fiefdom among many. What if the same creative process that builds a nation were devoted to its deconstruction?

Anthropologist Benedict Anderson's reflections on the origin and spread of nationalism, in his 1983 book *Imagined Communities,* helps to explain the creative process of drawing political boundaries. He asserts that a nation "is imagined because the members of even the smallest nation will never know most of their fellow members, meet them or even hear of them, yet in the minds of each lives the image of their communion."[3]

While nations may lean heavily upon myths of their rightful attachment to place and the kinship that binds them, each and every one relies on a singular leap of faith: Throw your lot in with "these" people as defined by "these" borders. Bingo, that is your nation. Instead of sifting through shared history to understand a nation, Anderson wants us to look forward and examine the creative processes that construct community.

Writing in the 1970s and 1980s, Anderson, Seton-Watson, and others were particularly interested in the creative process of nation-building because so many new nations were being born. The global mandate after World War II, brokered by the victors to shape the Bretton Woods world, liberated so many colonies.[4]

Or, at least, it set the process in motion. Over the subsequent decades, geographic boundaries, languages, and trade routes were tested as postcolonial leaders sought to build new states. With many boundaries drawn by the colonizers, new states often begat more new states. With each phase of construction or deconstruction, Anderson's reflections are especially relevant: If the nation itself is more a colonial construct than a historical precedent, then new leaders should be free to explore the creative process to construct new national communities. Writing mostly about Indonesia, Anderson understandably wonders how 17,000 islands can emerge from Dutch colonialism, and how can the inhabitants imagine a single nation?

Perhaps it is of little surprise that, during this period of frantic state-building, many new states embraced the model of single-party centralization. How better to impose and project these new imagined communities upon diverse populations? It is one thing to unite around the political project to overthrow foreign rule, yet quite another to govern as one people. State-building requires creativity.

Sekihara's kuni is a political project, devoted to pluralism. While it may not portend the dismantling of the nation state, the specific set of actions Sekihara calls for is akin to nation-building on a micro scale. Kuni is a creative defense of small places, an ambitious call for autonomy balanced with engagement with the rest of the world on its own terms. It is far better to trade on assets adored by outsiders, but curated by locals, than to allow the megacity to homogenize small vulnerable places like Nakonamata.

Japan emerged from World War II with one party dominating not only national elections but also the national imagination. The Liberal-Democratic Party has governed for most of the past seventy-five years. With the stability of one party, it is far easier to promote a single successful narrative of the postwar period: Bigger is better. Always! Big business, big government, and big cities deliver stability and security. To send this point home, greater Tokyo is home to thirty-nine million people, approximately 30 percent of the nation's population. This makes Tokyo what American geographer Mark Jefferson describes as a primate city. It is one that totally dominates power and public imagination for the rest of the nation.[5]

Maybe that is fine, if you are lucky enough to live in that primate city, but what if you live in one of the forgotten back channels outside Tokyo? Or in Sekihara's case, he grew up even farther away in a rural agricultural community north of Joetsu. While rice subsidies, economic growth, and light rail may have kept small communities like his in Niigata Prefecture on life support for years, any rural resident, with an eye for the future, is clearly focused on how to get to Tokyo. Sekihara is no exception.

How else could he have a successful career? The brain drain in small-town Japan made it all but certain: "Get the hell out of the village if you want a future!" This is true whether the village is in Niigata Prefecture or on the other side of Japan in Yamagata. I contend that the same is true of London, New York, and every other megacity that promises prosperity and a chance to ride the waves of opportunity.

Sekihara and I resumed our conversations six months later. We traveled to Yamagata Prefecture in northeast Japan to meet author, colleague, and Ehime University lecturer Ryoko Sato. Her writings about women and agriculture have influenced Japanese civil society and the subsequent Japanese farmers market phenomenon. At the dawn of the twenty-first century, her fascination with American farmers markets resulted in many trips to the United States and the publication of her book on the subject. Had she not been diagnosed with the terminal cancer that has since taken her life, she would have contributed chapters to this book.

In the spring of 2019 we were fortunate to confer with Ryoko Sato about kuni and sadly, say goodbye. During this trip, Ryoko introduced us to her father: a well-respected and published author and farmer. Tozaburo Sato is one the rural writers from his generation (age eighty at the time of this writing) described as "the peasant poets."

We met Tozaburo Sato in his farmhouse in rural Yamagata to listen to his insights for sustaining lives and livelihoods in rural Japan. Unsurprisingly, he holds a bleak vision of the future for rice-growing communities. He asks: "Who will keep the tradition of fermenting miso at the community level in the future if there is no community?" When prodded as to when the last time he remembered life was good for rural Japan, Sato responded without hesitation. "That's easy. The war. That was the last time rural goods and mountain knowledge were valued by the central government."[6]

Tozaburo Sato introduced us to another octogenarian poet, Michio Kimura. He and his poetry are the subject of Masaki Haramura's 2015 documentary *A Voiceless Cry,* about Kimura's activism, poetry, and defense of rural Japan. He too remembers World War II. That was when his father perished, an event that compelled him to pursue writing. After high school, he helped to form the farmer-poet collective that published an anthology called *Chikasui (Groundwater).* Today, clearly troubled by the unbridled global economy and the endless appetite for liberalization in trade, as the Trans-Pacific Partnership trade agreement has taken center stage for him, Kimura shared with us his disappointments: "We worked our entire lives to save village life and agriculture. You look around today and, despite the high points we may have achieved, I have to wonder, has it been worth it?"[7]

After an intense afternoon of discussion, pots of tea, and sweets, we left Kimura and Sato. I walked away rather despondent, wondering, Is it too late? Can anything actually save rural

Japan from its inevitable decline? Hilariously, by contrast, Sekihara left invigorated: "If those old men can still keep fighting, then what am I doing moaning and groaning?" If things have been this bad for so long in rural Japan, then no wonder it is time to embrace wild ideas.

Scale: Is There a Right Size?

If kuni (or community) is the goal, then how do we reach it? After losing himself in the brutality of big places (Tokyo), Sekihara returned to rural Joetsu. His return brought many surprises. Among the first lessons, and with the advantage of scale, he witnessed just how much rural maintenance has been deferred. How could he know? When he left as a young man, things seemed to work okay. Upon his return, not only could he see up close what deferred maintenance looks like, he was able to identify the responsible parties. He began to understand the tyranny of small places—with village mayors wielding authority, but with little accountability. Those who disapproved of the local regimes had either left or had grown despondent and unable to mount opposition. In response, Sekihara set out to address the neglected assets: canals and forests. In due time, these volunteer efforts morphed into a campaign to replace the village management of social, economic, and civic life with a new regional instrument—the RMO.

Sekihara uses the language of autonomy to describe the posture of the RMO. However, it is important to remember that the autonomy he seeks is not from the rest of the world or even the nation-state. Rather, he seeks autonomy from the small-minded and failing model of the village. Since kuni relies on the inflow of people, profit, and passion from city dwellers who forge lasting relationships during repeat visits, it may be better to describe his project as seeking interdependence.

Kuni is a Goldilocks quest for the right-sized community. Sekihara takes a provocative stab that the right size is between five hundred and two thousand inhabitants. Perhaps inspired by British anthropologist Robin Dunbar, Sekihara looks for balance. In the dying villages, Sekihara insists that there is not enough oxygen for pluralism to flourish. The village mayor rules with little opposition. Relationships are too close to risk taking on power. After all, if you lose, you have to live in close proximity to your enemies. As a result, the dictatorship of uniformity takes hold.

In Dunbar's 2010 book, *How Many Friends Does One Person Need?*, he determines that humans have the emotional and cognitive capacity to maintain no more than 150 stable relationships. He envisions bands of proximity that move out from the individual, growing larger in number with each leap to a wider and looser group. The smallest circle is family, at approximately 15 individuals. The largest group in which an individual can recognize others is 1,500.[8]

Dunbar's theory, known as Dunbar's number, is hotly debated: Is it universal? What about the internet? Is this true for both town and country? Within the context of analog community organizing, where trust and face-to-face relationships can make or break even the most elegant of plans, Dunbar's concepts ring true. Perhaps there is a sweet spot.

If too small a group, disagreements are difficult to accommodate. However, if too large, individuals lose their access to power. The caricature of when big is too big is the megacity[9] and why it feeds hopelessness. It comes down to this: no matter what I do, the machinery will go on without me, and in spite of me.

Sekihara has experience organizing villages of just a few dozen remaining residents. Lost and on their last breaths, he contends they are too small to survive alone. His innovation is the RMO that aggregates these forgotten villages, bypasses the remnants of leadership, and assembles a community closer to what Dunbar deems appropriate to cultivate familiarity and a shared purpose—1,500 people.

Feel free to dispute the scientific precision of whether 1,500 people is the right scale of social organization to achieve a desirable quality of life. After all, Dunbar's number has its critics. Regardless, the number has its precedents. Consider the secular utopian society in 1840s Monmouth County, New Jersey: the North American Phalanx. Organizers were devoted to French philosopher Charles Fourier's socialist theories of how to organize a self-regulating physical community. During its decade in existence, members sought to create a community of 1,620 people. Fourier determined that this was "the right size." The community's key buildings and businesses burned down in a fire in 1856. It was a short-lived experiment. However, its operations certainly occupy that sweet spot of scale to deliver balance between individual autonomy from and connectivity to others.

In the twenty-first century, the nation-state no longer maintains its hegemony as the tool of choice to deliver individual and collective security. Debate is fierce about where else to place this responsibility. While consensus evades us, many of the options skew large, if not larger, from international regimes to private-sector corporations. Meanwhile, others look smaller. The food movement has eyes on municipalities as an appropriate point of intervention. These are social organizations far larger than 500, or 1,500, or even Fourier's 1,620 people. However, I am encouraged that scale is on the menu.

Despite these questions of scale, our age remains devoted to it as a primary indicator of success. We are conditioned to accept the logic that bigger is better. No matter your game, you want an audience. Consider rock star Freddie Mercury. In 1985, journalist David Wigg asked him: "Does size of audience matter?"

Mercury responded, in predictable flamboyant form, with: "The bigger the better … in everything."

Is Mercury correct? Are things always better when bigger? Ask a group of people and many will agree. If you have something to say, you want to be heard far and wide. And yet, scale comes at a cost. Think of the classic music industry predicament. You can write and record music on your own equipment. Today, a laptop computer will do. Upload songs onto the web, and be your own record label, and broadcast to the world. Chances are, your audience will be limited. It is difficult to break through the din of competition in the marketplace without the gatekeepers who can deliver audiences in chunks. In order to gain greater distribution, you make compromises. If lucky, you get discovered by the artists and repertoire department (the famous A&R men) and sign with a record label. Now get ready to compromise. After all, those who shell out the money get to make decisions about both form and content.

While this metaphor may not dovetail precisely with the field of social change, the question of scale remains the same: The larger the project, perhaps the greater the impact, but also greater the risk for dilution, distortion, and waste.

In the world of politics and the relations between people and their governments, consider the calculations British voters took in 2016 to vote for or against the nation's membership to the European Union (EU). In a national plebiscite marred by all kinds of promises for those who vote to leave, a slim majority determined that the project of European governance had grown too large and clumsy to justify the perks that come with membership. By 2007, what began as a modest six-member alliance of Western European nations (circa 1951) had grown to twenty-seven, representing a population of five hundred million. Ever cautious to maintain a degree of detachment as an island, the UK did not join until 1973. However, Great Britain had been inside the community long enough for citizens to observe the growth. Had the political body grown too large and too concerned with growth to care for its members? After the fall of the Soviet bloc in 1991, the EU quickly absorbed nations that once formed the Iron Curtain. Putting aside the lies and xenophobic logic that fueled much of the campaign to exit the EU, an important question remains: What is the appropriate size of a community to deliver a healthy balance between proximity and anonymity, between precision and efficiency, and between the individual and the crowd? Do we find community globally? Or is the nation-state the place where we imagine belonging? Or, as Sekihara posits, is home smaller than that? This question strikes at the heart of Sekihara's efforts to reimagine his community, the rural territories located near, and technically within, a city of 200,000 residents—Joetsu.

I have devoted my professional life to championing the small farmer, beneath the shadow of the large. Operating from the geographic setting of the city, I reach out to rural. Is Sekihara my mirror image? Situated in the rural, he reaches out to urban. For me, the revival of the ancient farmers market became a primary instrument to link urban with rural people commercially, but also (and especially) socially. In the United States, the number of markets increased by 300 percent from 1996 to 2006.[10] There are a number of reasons for this mercurial rise; however, there is one rationale that fits closely with other social and political expressions, like the aforementioned campaign for Great Britain to exit the EU: People are growing tired of the scale of contemporary life.

In New Orleans, first-year architecture students are assigned to investigate architectural styles by visiting cemeteries. Why? It is simply easier to learn the different styles by examining them in miniature. A tomb is smaller than a house. You can walk up and touch the cornices. Similarly, I began to recognize the value in understanding the relationships in play in farmers markets. Weekly events that feature farmers, fishers, chefs, and shoppers may appear ad hoc but are, in fact, a codified social contract. Moreover, while the managing entity may not be center stage, someone enforces rules, promotes products, delivers programs, and projects an imagined community that includes a cast of characters.

Why does this informal-meets-formal activity work? Primarily, the surprise of seasonality and rejection of conformity differ greatly from the sterile environment of supermarkets, with their stock-keeping units (SKU) regime. Farmers markets are able to accommodate seasonality, individuality, and surprise because they are small. After all, when scaling up, it becomes more difficult to improvise. If the fresh scent and vibrant color of produce attract shoppers, it is the relationships that flourish in small social environments that keep people there. Farmers markets provide our glimpse into kuni's potential.

While it may be too soon to declare the age of globalization and its rush-to-scale over, the pandemic of 2020 has brought to the front burner many questions that have been brewing on the back ones for years. When the world grinds to a halt, we are afforded the time to ask.

2

KUNI IS COMMUNITY

Tsuyoshi Sekihara

A kuni is a community that emerges near a declining regional city in a rural area of a developed nation. A kuni is independent, but that does not mean it is independent from the nation. A kuni complements the local government.

What makes a kuni unique is that it is compact but contains all the elements needed for human life. It is hard to see the entire scope of kuni if you narrowly focus on nature, agriculture, commerce, a local organization, education, and welfare. Since a kuni is meant to benefit each individual in the community, it needs to reach the spiritual aspects of individuals. "Why am I the only one in this world? Why do others exist? Who am I, and who are others? What does it mean to live with others?" And, "When I live with others, what is the appropriate structure and size for humanity to be most valued?" These questions can't be explained in a concise way, like a how-to manual. But that is where the key to kuni lies. The key is to know the appropriate size. That means it is neither too big nor too small. A medium-size piece of land, a medium-size population, a local organization of medium size, a local industry of medium size for economic independence, moderate scale of activities for local people, and a moderate number of urban people visiting the kuni. A medium size so people can live as humanely as possible. Pursuing "a right-sized community" may lead to more possibilities in the future. It is hard to imagine what a right-sized community is in a developed nation.

The common thread in the following chapters of this book is identifying the right-sized community. I would like to explain this through my modest experience. It is not epic, but rather just a small story that I hope you enjoy.

3

THE DAY IT STARTED

Tsuyoshi Sekihara

It was the beginning of winter in 1994, and I was in a small park in a suburb of Tokyo. I was thirty-three years old. The sky was clear and the sun filtered through the trees on a path covered with fallen leaves. It was a weekday afternoon and no one else was there. I stopped in front of a tree and picked up a rope. It was probably left behind by a gardener who was doing prep work for the winter. I looked at the rope for a while, and tied it around a branch, and stretched it to hang myself. But the rope snapped easily and I landed on the ground.

This experience led me to start thinking about kuni.

A small company that I co-owned went into bankruptcy. Something like this happens a lot and would not draw any attention. However, it can be a life-or-death matter for the individuals involved. In my case, the rope snapped, and I decided not to try it again. At the same time, I realized that I was a failure, and I was still alive.

At dusk that day, I roamed aimlessly around the city center's narrow alleys. I ended up in an old and shabby public housing area. At twilight, the windows were lit, the smell of dinner wafted out, and I could hear the laughter of the children. I will never forget the warmth of the lights, the rich smell of food, and the wholesome laughter.

I felt as if a deep fog had just lifted and cleared my sight. I was the one who created the fog and blocked my vision. The fog that blocked my view was the socially acceptable values at that time.

People were not deciding for themselves what constituted success or failure. We just believed in what schools, companies, and the media were telling us—in other words, what the system endorsed. The fog that blocked my vision didn't allow me to question the system in which I lived. We allowed the system to determine success and failure and did not think about what this meant for ourselves. We did not even question if success or failure existed. Although I had a small success, I bought into the idea that the more I could afford determined my success. It was a way to win in the game of life, but it didn't say anything about the value of life itself.

It was pathetic that the deep fog I was in lifted only at the crossroads of life and death. But once the fog lifted, the world was fresh. All the old values that dictated my life shrank, and hidden values started to shine through. A clear vision was something I acquired only after a horribly big setback.

It would have been impossible for me to lift the fog on my own. It required an incident, an accident, a disaster, an illness, an unexpected setback. That is why I want to tell young readers that there is meaning in setbacks. The system of winners and losers that society teaches us is designed to create a handful of winners and many losers. In fact, most young people are forced to learn about losing by coming across the winners of their age on the internet. Then you have to learn how to come to terms with being on the losing end. As long as the world is dictated by that game, no one can get away from the pain of losing. In that sense, there is value in encountering major setbacks. A setback is a curse for someone who would like to stay in the game, but it is a blessing for someone who wants to leave the game.

Therefore, if you experience a setback, just deal with it. It is painful, but if you deal with it thoroughly, you can look at yourself objectively as someone who was dictated to by "the game." In addition, setbacks help a person to understand solitude fully. It is important to recognize that you are alone. By fully grasping the notion of solitude, you are able to understand others. When your grasp of solitude is incomplete, you use and consume others as a tool to alleviate your loneliness. This is narcissistic behavior. Others do not essentially exist for this person.

The core thinking of kuni lies in the willingness to form a community. In other words, it is the will of wanting to live with others. To do this, you have to fully understand what solitude is and feel for others viscerally.

I want to let young readers facing setbacks know that it is precious to experience them when you are young. This is something that I experienced.

There was another big factor that led me into thinking about kuni. On January 17, 1995, two months after the rope snapped in the park, the Great Hanshin Earthquake occurred. The footage I saw looked staged. It was hard to believe a big city could crumble, as if building blocks were being knocked down.

After I returned to the place I was born, it also was struck by an unprecedented huge earthquake on October 23, 2004. Landslides from the mountains squashed houses. The roads were twisted as if they had turned into roller coasters, and utility hole covers protruded from the road as if they were some strange artwork. Then on March 11, 2011, the Great East Japan Earthquake struck Tohoku. I felt a long horizontal sway that made me nauseous. I turned on the television to discover unbelievable footage. I quietly watched the giant tsunami swallowing towns one after the other. It looked like a film. In a span of sixteen years, four big earthquakes revealed the fragility of urban areas. The land on which people built, remodeled, and dominated the landscape was easily torn apart like a doll made from paper. The events destroyed buildings and people's lives.

These cascading experiences made me realize the importance of the power of the land. It was only human hubris to think that we could control and dominate it. No one can contain the power of the land—the power will always stay there. That is why it is important to acknowledge the power of the land. It allows us to grow crops, yet it is also where disaster happens. This is common sense, but most of us—including myself—have forgotten.

The Place Where It Started

In March 1995, two months after the Great Hanshin Earthquake, I moved to Joetsu City, Niigata Prefecture. I grew up with nature in a rural area, so I didn't feel anything special when I looked at the mountains and the rivers. Even though "the fog had lifted," I still felt a sense of failure and loss. One day, I was driving my car aimlessly on an unfamiliar mountain road. I drove through many dark ridges soaked with light rain, but I did not encounter any inhabitants. When I was about to turn back, I suddenly found a village at the bottom of the valley. The scenery in front of me made me feel as if I had traveled back a hundred years.

It was raining in the village, so no one was outside. I kept on driving the narrow road. The village was bigger than I'd initially thought. There were many houses. At that time, I could not be bothered to show interest in the history or the life of the village. I was overwhelmed with a baseless sense of fear.

I thought that maybe the residents were staring at me, an outsider, quietly from inside their houses. I was afraid that a group of them with sickles in their hands would banish me from the village. In hindsight, it was stupid paranoia, but that was what I felt back then.

This village became a cultural symbol for its richness in survival skills and was called "The Heart" of villages we worked in after we established a Regional Management Organization (RMO) in the area. Many concepts that are the basis of the theoretical structure of kuni came from observing this village.

I don't think my senseless fear was completely off the mark. While I didn't really think the villagers were going to come after me with sickles, insular villages are subconsciously unwelcoming to outsiders. An outsider like me is considered a traveler *(tabi-no-mono)*. Even if I had lived in the village for ten years, I would be considered a traveler, a status that would remain for thirty years or more. You are known as a traveler for the rest of your life. When I die and my descendants take over, the title would finally be removed. It's not that you experience discrimination because you are a traveler, but you are subject to distinction.

After I drove through the village with a little bit of dread, I entered a mountain road to cross a ridge. The winding narrow road continued. Once I crossed the ridge, I arrived at a bright and open space along a river. The rain stopped and the place was filled with soft sunlight.

I drove down the road along the river. There were houses, one after the other. I was surprised to learn that a village of this size had existed in a remote area for such a long time. It was a valley, but the mountains were low and sunny. The gardens in front of the houses were filled with flowers, and every time I rounded a curve I'd come to a new village. I was enchanted with what I found.

The flowers were beautiful. Many years prior, I had visited New Zealand, where I was enthralled with a decorative garden. Flowers and shrubs were selected as a feast for the eyes. They were meant to impress. By contrast, here in the village, the home gardens were not showy. At first, I could not grasp what was different about these beautiful flowers from the ones that stuck with me from my experience in New Zealand. Upon closer inspection, I understood. These flowers come from fruits and vegetables. They were planted for survival. Not all; some flowers were there for beauty, but they existed in balance with everything that was planted for consumption. Nothing was there to make a statement to others. They planted everything for themselves—to look at and to consume. They were not designing a garden. They just planted what they needed. I was simply lucky enough to see the outcome of their design and their labor. This lack of pretense actually made the trees and flowers look more beautiful.

In Japanese, there are two words: *makanau* (賄う, "self-sufficiency") and *tsukurou* (繕う, "to use something by constantly fixing it"). The *u* at the end indicates continuity in Japanese. The continuity of the hamlets that I saw was formed with makanau and tsukurou. Their beauty is not something decorative. Their beauty derives from practicality. Another Japanese word, *younobi* (用の美, "functional beauty"), also helps to describe my experience. The warmth that the hamlets exuded was similar to the lights that I saw in the public housing in Tokyo on the day that the rope snapped: human life, serene and plain.

After my return from Tokyo, I saw the mountain and the sea in monochrome. Since that day in the villages, I began to see colors again. My rusty five senses were revived. My hope was revived too. I found a place that was not dictated by winning or losing. To live for living is all that matters. That discovery gave me hope. Several years later, these villages deep in the mountains, along the river, and along the coast became the base of operations for what became our RMO. A total of twenty-five villages with a population of 1,900 became my new world and our shared purpose for living. Back then, it was something that I could not even imagine.

Discovery of Shadow

After that day, I started to visit the villages frequently to enjoy the beauty in harmony. With each return, I began to see things that had been invisible to me in the beginning: vacant structures, paddy fields that were no longer tended, forests that were abandoned, and many elders and few children. I learned that the local government limited the construction of new houses. (In other words, they wanted the residents to leave the village and move to urban areas.) Public transportation service was decreasing, and there were only two roundtrip buses a day. Young people hated the winter with its heavy snow, and they moved to the city. (Their workplace was in the urban area and not in the village.) The village was quietly heading toward extinction.

When I spoke to the residents, they described their love for their villages. Although worried about the decline, no one had any ideas on how to stop it. In fact, they were the ones who told their children to leave the village and go to the city, as there was no future in the village. We can't blame them for saying that. They pondered the future of their children and thought that was the best option.

It is hard to tell when, precisely, the residents in rural areas started to give up. After the postwar economic growth, cities became the developed places and rural areas became backward places. Staying in a rural area meant defeat, while living in a city meant success. The media repeatedly ridiculed rural areas as muddy places, smelly with manure, and filled with people who speak with dialects. As this pattern repeated, urban areas reinforced the collective illusion: sympathy without reason. A lot of urbanites would say they felt badly about rural areas, but without any clear reason.

"If It Is That Great, Why Don't You Just Live Here?"

Let me go back in time to 1977. I was a high school student, conducting folkloric research in a village. All of the residents decided to abandon the village and move somewhere else. I came across a magnificent *kominka* (a traditional house). It was probably built more than two hundred years ago. While conducting research on the *kominka,* I asked the woman who married into the family that owned it why they were abandoning such an incredible place. She replied, "I have the right to live in a Western-style house with flooring, white walls, and a white kitchen." I could not argue with her. White walls, a white kitchen, and a white refrigerator. There was nothing white in this village. It was full of dying leaves.

Villages Killed by Glorification

The decline of rural areas proceeded over a long time, almost as if it had been planned that way. Pretending to sympathize, glorify, and protect rural areas, people undermined the vitality of the villages. The ruling party took advantage of the decline of the rural areas, and under the guise of sympathy, poured in huge amounts of subsidies. The opposition parties praised and portrayed rural areas as utopian and an ideal place for their ideology. Both parties glorified rural areas for their own purposes, but they did nothing to secure the future of rural areas. As a result, three hundred villages disappear every year in Japan.

In recent years, other types of people began to exploit rural areas. There are people who preach with an emotional tone about the misery of rural areas as victims, while others praise villages as paradise. These activists, demagogues, artists, scholars, and politicians treat rural areas as if they are toys. Worse, they manage to make money off these relations. Perhaps, this is why they attack and silence opinions and criticisms that rural residents need to hear in order to forge a future. Turning meaningful discourse about rural liabilities into a taboo subject, they interfere with necessary change.

Urbanites may wonder, "Why have rural areas declined despite all the subsidies they received?" The answer is simple. Since rural areas were saturated with subsidies without strategy or tactics, their vitality was lost. In other words, urban centralization policies planned by central and local government succeeded. When I returned to where I came from, I saw the possibility of a new community born out of the power of the land and farming in mountain villages that were otherwise heading straight, yet quietly, into extinction.

4

IT BEGAN WITH A FOREST

Tsuyoshi Sekihara

In 1997, two years after I returned home, I got a job as the executive director of a woodworking cooperative. The executive director wanted to retire and asked me if I was interested in the position. I asked if I could go home at 5 p.m. every day as I wanted to go fishing. He agreed. And it was a done deal. But I had no time for fishing.

The woodworking cooperative led to the establishment of a forest NPO, a cross-sector cooperative for the timber industry, and eventually an RMO. The cooperative was made up of companies that primarily produced wooden doors and shoji screens. The long recession that began in 1990 led to a sharp decline in construction work in regional cities. In my region, the number of wood product companies dropped from 150 to less than 50; they were either going bankrupt or closing down. The cooperative was created to do something about that situation.

The first thing I did at the cooperative was to point out that there was no future if we only made wooden doors. Since we had carpentry skills, we had to expand the range of wooden products to more than doors. The people I worked with were manufacturers of Japanese-style doors and were skilled in processing conifers (cedar and cypress). In those days, wooden furniture was mainly made from broadleaf trees and very few were made from conifers. We decided to build furniture using conifers, which was our specialty. If we built furniture, we thought we could sell it directly to customers. Having said that, wooden furniture was available everywhere. We decided that we would not only use conifers, but conifers from the local forests, which were cut to care for and thin the forest. The devastation of conifer forests was already becoming an issue, and we wanted to create added value by stating that the furniture helped preserve the forests.

Sacrifice for Export

The reason forestry in Japan declined is simple. Roughly speaking, it is due to the rise in Japan's economic power and the appreciation of the yen (the Japanese currency). That made imported timber half the price of domestic timber. Furthermore, the government tacitly approved the import of large quantities of raw timber from countries where they wanted to export industrial products. This meant sacrificing domestic forestry for economic growth. Small-scale forestry could no longer survive. The domestic forests that were sacrificed were not small. Japan's forests, as a percentage of its total landmass, were ranked third in the developed world, and half of its forests were planted. Many people think that Japan is a small island, but Japan is bigger than England, Italy, and Germany. In Europe, Sweden, Spain, and France are the only countries that are bigger than Japan. The forests that were sacrificed were not small.

Eventually, instead of just sacrificing domestic forestry to strengthen exports, the government compensated affected regions with huge subsidies. It was the same with agriculture and was also a way to buy votes for the ruling party. The subsidies were effective, and people working in forestry came to think that it was better to abandon forests as victims of an export-based economy than to fight against a policy that sacrificed forestry. Under those circumstances, when the cooperative tried to purchase local timber from a wholesaler, no one had stock. Where was the local timber? Trees were growing in the forests and were on the brink of ruin. No company would cut down uneven trees that were unsuitable for mass production. Therefore, there was no timber in stock. This later became an important lesson that informed my thoughts on how an RMO could help reinvigorate an industry using local resources. If you want to start an industry using local resources, it is important to create a structure that includes both extraction and an inventory of local resources. You have to be involved with the entire process, from growing to turning the local resource into a product. When you are creating a local industry, the product has to be unique and have high added value. However, to get there, you have to first find ways to transform available raw materials into useful materials.

Stepping Out of the Existing Distribution System

Developing a small local industry using local resources means you have to leave the existing distribution system. This was shocking to us, as we believed that we could buy any material as long as we paid for it. We were lucky because we figured out early on that we needed to get out of the existing distribution system and create our own small distribution system. This thinking was not only useful for the cooperative but also for the regional organization. We wanted to create a wide variety of local industries that were independent economically. If you want a local industry with small output and diverse products to thrive, you have to go outside the existing market and the existing distribution system.

Crayons with Six Colors or Twenty-four Colors

In order to get timber, the cooperative decided to purchase trees growing in the mountains. We paid for the trees, including the cost of cutting them down and sawing them. We somehow managed to stock local timber, but the quality was uneven. This was because no one had taken care of the forest for many years. Furniture makers expressed their discontent, and I told them: "The timber that you usually purchase is all the same and lacks in individuality. In other words, it's like a crayon set with six colors. However, local timber is like a crayon set with twenty-four colors. Depending on your skills, you can add value to it."

This may sound desperate, but I was not lying. Individuality makes low-output products all the more attractive. I also started to see another outcome as a result of this. The furniture makers became more skillful, as they had to creatively tackle timber of uneven quality. If you only work with the same material, your skills deteriorate. If you use material that is difficult to deal with, it improves your skills. For mass production, it is essential that material is standardized to fit the machinery that makes mass production possible. Therefore, uneven material is not used, and carpentry skills are not incorporated into the process. We saw an opportunity for survival by finding value in something that was not meant to be used in the first place. Unfortunately, we also realized that this opportunity would vanish due to events that ran counter to our intention.

Around this time, we saw an increase in the closure of small sawmills in the area. The government's direction was to integrate all the small and inefficient sawmills into one major sawmill for the region. As a result, a huge sawmill was built with government subsidies, but it went bankrupt several years later due to excessive investment in equipment and an inability to meet even half the sawmill's production capacity. This coincided with a massive amount of cheap mass-produced products from a bigger factory in another region flooding the distribution system. The added value gained through capital investment in machinery crumbled the moment a competitor installed the same machine. That is the extent of the added value that machines can bring versus what carpentry skills and intelligence can bring.

This is what ends up happening when the government guides regional industries. The greatest sin they committed was not the bankruptcy of the big company, but the closures of small sawmills in the region. In order to have a small-scale yet high value-added timber industry, we needed fifty small destroyers instead of one large aircraft carrier. The choice the government made led to defeat in the big market and diminished the possibility of the small market. Many of these small markets are in the public architecture sector in regional cities: city halls, schools, community centers, libraries, hospitals, stalls to sell local produce directly, tourism offices, and wooden signs. If the government wanted to strengthen the small market, there were many possibilities.

Something We Can't Do

The government made another mistake. They would not change their rigid approach where it mattered most: the bidding process.

In other words, even if the furniture was made by local people using local timber and had the same function as mass-produced products, if the bidding price for mass-produced items was lower by even one yen, local industry would lose out to mass-produced steel furniture.

Wine Label

Despite these circumstances, we moved forward with our projects at the woodworking cooperative. The next issue that came up was how to certify the local timber that we purchased with painstaking effort. I see the forest as a field and the timber as the agricultural product. I thought we could put a label on a wooden chair in much the way we label wine. The wine label shows the grape varieties, the vineyards where the grapes are grown, the winery, the technicians, and the philosophy on wine-making. With this in mind, we decided to label the wooden chairs like wine. The problem was figuring out who could certify the place of origin. Trust is everything in a certification process. Someone who profits from sales is not suitable. We decided to create an independent forest nonprofit organization. This was back in 1998. I believe this is one of the very first NPOs in Japan that dealt with forest preservation. We adopted a local certification process, which was primitive, but it meant that someone from the NPO needed to go to the forest directly to verify the timber. With this methodology, you can only cover a small area, but a thirty-mile radius was enough. If we needed to expand our efforts, all we had to do was to establish another NPO in that area. I still think this is the best certification process if you want to nurture a local timber industry in each region.

The cooperative started to purchase certification stickers from the NPO, which stated the place of origin. It cost five hundred yen per sticker. It was a small industry, but we paid more than one million yen to the NPO in profitable years. Later, the cooperative also received Forest Stewardship Council (FSC) certification to deal with big markets, but the process was too costly, and the cooperative left the FSC several years later. The FSC was suitable for companies with mass-production capabilities, but not for small companies. At least I learned something. The FSC dealt with a massive amount of certification and needed to rely on documents and vouchers, and this led to some false certifications. It could not match the primitive yet precise certification process of a small NPO that could go directly to the forest. The NPO, which worked as a partner with the small local industry, was also small, but it was reliable. A globalized certification process is effective for products that are distributed globally. In this instance, "one size fits all" does not work, and the global scheme was not necessarily going to be effective in small areas.

This is not only about woodwork. How do you certify products from a small farmer's paddy fields, a grandmother's vegetable field, or a fish haul from a tiny port? For small local brands, it is necessary for them to have a small yet reliable local certification process. It can be implemented, and it is not too difficult. A small group of residents could work with local industry. There is nothing to worry about if everything is certified locally, because there is no room for false certification. If this were to happen, the producer and the local NPO staff would all need to lie together. It would be too cumbersome to do something like that. The cooperative and the forest NPO I was involved in are still in existence and, although on a small scale, continue their certification process.

These efforts at the woodworking cooperative taught me important lessons when I later founded a Regional Management Organization. For the RMO to be independent economically, a local industry that uses local resources is essential. My experience running the cooperative helped me run the RMO. The management structure of an incorporated NPO and a cooperative is similar. Both have bylaws, members, a board of directors, and an administrative office. Decisions are made at board meetings and the annual gathering. Everything I did prior to founding the RMO was essentially basic training for running the RMO. I also specialized in timber, which led me to the ability to "bricolage," in other words, to make something beautiful with material that exists in front of you. The path from the woodworking cooperative to the forest NPO, and then to the RMO, was not planned. The path appeared as I solved one problem after the other. Nothing happened by chance. When you look for ways to solve problems, you are also seeing all the causes that lead to those problems.

In 1999 the woodworking cooperative and the forest NPO received awards from the government for their efforts. There was even a program on the public broadcasting network. The staff were happy, but I didn't care. What I wanted from the government was not a medal, but a market. Two of our wooden benches are placed at the main entrance of the headquarters of Japan's Ministry of Agriculture, Forestry and Fisheries. For them, buying the benches was a way to reward us. The benches just look like sad objects to me.

5

RURAL CRISIS: The Last Straw

Richard McCarthy

When Masanobu Fukuoka published his influential agroecology manifesto, *The One-Straw Revolution: An Introduction to Natural Farming,* in 1975, Japan's economic miracle was in full bloom. So too was the great migration to Japanese cities. While workers were throwing themselves into long commutes and corporate culture, Fukuoka provided an alternative perspective on work: "Human beings are the only animals who have to work, and I think that is the most ridiculous thing in the world."[1] While he could not have situated himself further from the zeitgeist of post–World War II Japan, Fukuoka inspired a back-to-the-land movement with methods like no-till agriculture and seed-saving at its core.

If you have read Fukuoka's books or watched him on film, you can understand why his influence has been felt far beyond farming. His critique of civilization and modern ideas about work and pleasure place him squarely with the likes of Rudolf Steiner, Carlo Petrini, and Wendell Berry, who wrote the preface to the US edition to *The One-Straw Revolution.*

When I asked Sekihara about Fukuoka's ideas and whether he influenced his Kamiechigo Yamazato Fan Club, he nodded enthusiastically and said, "Of course." This response is a clear example of how fully Fukuoka's work has been felt throughout Japan. It rests in the public imagination as a conversation starter. Fukuoka's work addresses the scientific validity of natural agricultural practices and offers techniques to calibrate our lives into the seasons and pace of the planet. But he proposes few specifics as how best to operationalize his wild ideas into practical solutions at the scale of rural development.

I contend Sekihara is an inheritor of Fukuoka's "One-Straw Revolution." He takes it one step closer to practical application. Fukuoka's brilliance is that he sees how "a revolution can begin from this one strand of straw."[2] But what if the straws disappear? Grains, and especially rice, are essential to the Japanese landscape, flavor, and imagination. While national agricultural policies may continue to protect indigenous rice production (to some degree), there is no escaping the cumulative effects of depleted rural communities and economies. If one straw can start the revolution, the reality in rural Japan (and much of the rest of the world, for that matter) is that few remain. If we do not act soon, it may be too late.

The news media regularly reports on the decline of traditional Japanese life as an inevitability. One example from 2019 is from an article in *The Economist* titled "Brides for Bumpkins."[3] The story reveals the lengths that rural governments will go to lure urban singles via a state-owned digital portal for matchmaking. While the myth of village life remains vital to what it means to be Japanese, in reality very few people know their way around terraced rice fields today. What lengths will they go to in order to repopulate rural Japan?

Elected officials are painfully aware of this crisis. Political parties struggle to find candidates to run for office, let alone attract voters to go to the polls. Rural residents live this decline daily: With too few children to attend class, schools are shuttered. Social services are also withdrawn; municipalities are being unincorporated, merged, and generally left to fade into oblivion.

Sekihara describes this crisis as the spark that gave birth to his Fan Club in 2002. At the time, he could not have anticipated that his modest community forestry intervention would spark a new kind of intermediary organization: a Regional Management Organization. He took the personal risk to repair and manage the forests in his new adopted home of Nakanomata. This was the Fan Club's birth.

Sekihara survived Tokyo's frantic confines for a decade. Its hustle and bustle captivated his imagination while growing up in another part of rural Niigata Prefecture, about an hour's drive north of Joetsu. Located twenty minutes south of Joetsu City, Nakanomata and its surrounding environs spoke to Sekihara so profoundly that they almost serve as the protagonists that set him on the path to build the RMO. With the natural beauty of the forests reminding him of the forests of his youth, key elements had changed for both the region and for him. Sekihara had hoped to find solace in the forest. Instead, he discovered abandonment.

Did this mirror how he felt about the trajectory of his life and career? Those who have devoted their lives to social change may recognize how difficult it is to separate the personal from the political. Does it take an emotionally damaged individual to dive into emotionally damaged communities?

On a practical level, who cares for rural infrastructure? Sekihara describes how horrified he was to find irrigation streams clogged with debris. The region is famous for its thousand-year heritage of rice cultivation reliant on communal irrigation systems. Niigata rice has long been coveted by sake aficionados for its cleanliness, resulting from the deep snowpacks that accumulate in the mountains each winter. Each snowpack serves as an almost endless source of clean water for rice cultivation.

In addition to agriculture, inhabitants foraged and hunted for valuable resources to survive: firewood for fuel; salt, mushrooms, wasabi roots, and game for food. In order to rely on these available resources, locals took it upon themselves to maintain these natural assets. With modernization, local governments began to absorb responsibilities once organized by the community. As people left for the cities in search of work in the 1950s, so too left the tax dollars needed to underwrite the public management of the forests. What has transpired since is a slow unraveling of natural resource management.

Instead of finding carefully groomed forests, Sekihara encountered poorly maintained watersheds, thickets of small trees that should be thinned for kindling and to prevent fires, and debris dumped by public- and private-sector scofflaws. The state of the forest was Sekihara's first clue that no one appeared to be concerned with maintaining the well-being of life in the region. These early observations led to well-being becoming an important tenet of his RMO's work.

The deferred maintenance of the forest became Sekihara's unexpected opening to shape life in his newly adopted community. Clearing irrigation streams, at first alone, and then with neighbors, became the spark. When he asked local mayors for assistance, they responded with indifference. It is as if they had acquiesced to the winding down of local services and its subsequent result: a loss of local pride.

Undeterred, Sekihara organized volunteer brigades of locals with shovels, ladders, and hoes to trim the trees and to dig out silt from irrigation canals and streams. These volunteer efforts started small but grew to 4,500 hours of service each year. More importantly, Sekihara and his neighbors stemmed the flow of apparently inevitable decline.

The Great Hollowing-Out of Rural Is Global

By no means is Japan alone in navigating this great hollowing-out. Powder kegs for populism in some places and home to drug addiction and despair in others, each community experiences loss differently. However, everywhere, cities rule. In the United States, President Trump swept to power in 2016 by claiming he would fight for white rural America. Irrespective of Trump's actual intentions or the follow-through of his policies, he recognized untapped potential for political mobilization and manipulation. Why did this success surprise the political establishment?

Aside from the cultural relics of pickup trucks, NASCAR, beards, and banjos that cling to advertisements as indicators of authenticity, rural life no longer dominates the public imagination or public policy priorities. Cities rule. Consider private philanthropy as a proxy for American priorities. According to US Department of Agriculture research from 2010 to 2014, only 6 to 7 percent of private foundation grantmaking was devoted to rural areas.[4] Where persistent rural poverty dominates, the contrasts are even starker. Writing in the Harvard Business School's Institute for Strategy and Competitiveness online publication, *Reimagining Social Change,* philanthropy executive Wynn Rosser points out: "The Alabama Black Belt and Mississippi Delta receive $41 per capita compared to New York City's $1,966 per capita."[5]

Of course, decline tends to be just that: a slow downward slope. Natural and unnatural disasters can accelerate decline with the dramatic effects of floods, fires, and firings, but until recently, they have been rare events. In the United States, the "get big or get out" farm crisis of the 1980s revealed hidden weaknesses in rural economies and an overdependence on commodity agriculture. As a result, the many independent institutions that once kept the circulation of rural money touching many different hands disappeared, giving way to consolidation—from the grain elevator operator to the local bank. I discussed this gutting of rural institutions in Mississippi with Ben Burkett, an influential African American farmer. As he describes it:

> We used to farm a lot more land for soybeans and other commodities. We would bring them to the locally owned elevator. It's gone. So too is the feed store. The only things left are the institutions we own collectively—the church, the vegetable processing shed, a delivery truck, and some farm equipment. Had we not made that effort to join forces as a community, we'd be gone too.[6]

By the 1990s, the long hunt for efficiencies and industrialization had left a permanent dent on American agriculture and rural communities. In 1935, a once pastoral nation boasted 6.8 million farms.[7] By the 1970s the shift to industrial agriculture was complete, leaving 2.2 million farms in production.[8] Since then, the number of farmers has continued to fall, but by then most of the damage on family agriculture was done. In 2020, USDA reports there are 2.02 million farms.[9]

The steady race to scale and monocrop agriculture has helped to create a bifurcation where there are those who operate in the grid of commodities and subsidies, and then there is everybody else. Many small- and medium-size farms folded beneath their weight of debt. Some grew huge, gobbling up land available at bargain prices. Meanwhile, others had grown comparatively small and so desperate for new revenue streams that they turned to antiquated, yet unexpectedly fruitful, business strategies like direct marketing and agrotourism. While this decline in American small-town Main Streets and anchor institutions like independent banks and farms continues, the change is far more dramatic in Japan. To make matters worse, options to regenerate rural communities hemorrhaging people and resources are limited by Japan's extremely strict immigration policies. They deny shrinking populations any potential for replacements from around the world.

Boom, Bust, and Then the Existential Crisis

Travel beyond Japan's familiar tourist destinations to rural Japan, and rather quickly, similarities to other nations are evident. This is especially true when compared to places where the physical and political infrastructure changed dramatically after World War II. While Japan's remarkable rail and road network may take you to "charming" fishing and rice villages, prepare to be underwhelmed. Far too little remains. Life is reserved for the big towns. Rapid industrialization, economic growth, and a commitment to civil peace between labor, management, and government created a stable democratic pathway to prosperity. Unfortunately, Japan's postwar corporatist consensus only envisaged a future in large cities—that is, until the technological and political disruptions of the 1990s changed the rules and sent Japan's economic miracle into a tailspin.

Think of Germany, France, and Great Britain for comparisons. These nations enjoyed exceptional economic growth during the postwar years, based on modernization projects and a concentration of industrial output aggregated in cities. With the end of the Cold War, Germany was reunited and invested heavily in the former East Germany. Japan's low-cost yet high-technology copycat economy of watches and televisions in the 1980s left the nation unprepared for the information technology economy that sprang from Silicon Valley in the following decade.

I think back to my first visit to rural Joetsu. Our minivan careened around the winding roads back down to the coast and to the town proper. Being driven on the left side of the road, I could not help but ponder this strange artifact of left-side drive. Most countries that drive on the left (only about 35 percent of countries[10]) do so because they held some sort of close relationship with Great Britain, but in Japan the rationale has no such connection. Rather, it harks back to Edo Period samurai, who preferred to carry their sword in their right hand when traveling. With paths almost always narrow, samurai would encounter others traveling the opposite direction in close quarters and with swords always ready for action. This wonderfully archaic holdover is so reminiscent of how and why traditions remain in both Japan and Great Britain.

While no two countries are identical, comparisons between Japan and Great Britain are particularly compelling. Both island nations once accumulated vast wealth via large mercantile empires. Cars are driven on the left side. Both systems value a strict adherence to protocol, social hierarchy, and celebrated constitutional monarchies. And there's more: Whereas the US political system distributes power via a federal system, both Japan and Britain concentrate power in central governments. (It is worth noting that beginning in the 1990s, Britain began to unravel some authority to regional national assemblies, like Scotland.) Nevertheless, the postwar period has enabled central governments in Japan and Britain to accelerate social and economic change rapidly. Unfortunately, this has steered much of the growth to cultural, financial, and political centers in Tokyo and London. Financial institutions congregate in these capital cities, but so do intellectual hubs: universities, think tanks, museums, and corporations, among others. As a result, small communities struggle to be heard.

By contrast, in North America, the federal diffusion of power serves as an internal regulator to change. In angry times, locals may depict the federal government negatively, hurling insults at the "do-nothing Congress" or "government gridlock." In happier times, it may help explain why rural decline in the United States (and Canada, also a federal system) has moved at a slower pace. As an example, the intellectual capital assembled at US land grant universities exists all over the country, not just in large population centers. Similarly, it may also help explain why North American responses to this decline are slower, and muted: "Crisis? What crisis?" The existential threat to rural life in North America seems urgent, but less urgent. Nevertheless, during Sekihara's tours of the United States, it was apparent that rural advocates thirst for his language and holistic approach to community organizing. Sekihara very much felt at home in Nebraska when visiting the Center for Rural Affairs.

Creative Responses to Rural Decline: Tourism and the Squad

The hospitality industry is often a beacon call for low wages and the development of policies that prioritize the needs of visitors above locals. It need not be this way. Two noteworthy strategies address the rural appetite for new money and new people by directing urban people to rural areas: Tourism and a civilian corps, referred to as the Squad. While these strategies are united as one in communities like Sekihara's, to do so requires intentionality. It is far easier to allow the tourism industry to impose its cookie-cutter models of development. However, Sekihara posits an asset-based strategy that has less to do with mass tourism and more to do with community development.

Tourism

In Japan, urban dwellers yearn for the very indicators of life that are missing in the big city: tradition, seasonality, and the kind of peace and quiet enjoyed in the forest. Ask dwellers in North American cities and they may also articulate similar deficits in their lives. Without question, these qualities are valued in Japan. Think of the place that forest bathing has in Japanese life.[11] With Japan's pandemic-plagued quest to host the Summer Olympics, originally scheduled for 2020, elected officials began to leverage the anticipated gathering of sports fans for the Olympics as a logic for investing in mass tourism in rural areas, easily accessible by Japan's remarkable rail system.[12] Unfortunately, mass tourism fails to deliver the quick fix for local communities unprepared to welcome visitors. Money may show up but not stick around if visitors encounter insufficient services. They take time to assemble. If visitors cannot find adequate hotels and restaurants, they are likely to leave without spending much cash, or worse, decide not to come at all.

The quick response to these local shortcomings is to lure in hotel chains and fast food to "give the people what they want." True, these methods may provide some employment: first to build new facilities; and second, to provide service jobs for the long haul. And while it may be a safe gamble to deliver familiar services, it does little to cultivate local entrepreneurship that generates wealth. Perhaps, just as important, it does little to distinguish one destination from another. Visitors may come, but then wonder why they bothered. Isn't it like everywhere else? If, instead, resources are deployed to generate a different kind of tourism, to attract a different type of visitor, maybe this will help to shift the narrative to assets, as opposed to the long list of deficits that plague rural places.

In this regard, Sekihara's strategy to attract repeat visitors deviates from run-of-the-mill mass tourism strategies. You will learn more in subsequent chapters about the Rice Covenant. When a consumer purchases a bag of rice, the transaction brings with it membership. The Rice Covenant is an invitation for urban dwellers to join the rural community, invest in it through voluntourism, and reap the benefits of belonging. One benefit of belonging is the right to attend community rituals otherwise open only to locals. Another is an insurance policy: the guarantee that you will find safe haven away from disaster-prone cities during times of crisis.

The Squad Program

The Squad is a service corps program, much like AmeriCorps in the United States. Its full-name is Chiikiokoshi Kyoryoku-tai ("Community-Reactivating Cooperator Squad"). It was introduced in 2008 during a rare moment in modern Japanese political history: The opposition parties rose to power in a historic election. By resettling young urban migrants into rural communities desperate for youth, most Squad members add value to agriculture and learn skills with the intention of remaining as permanent transplants, won over by all that rural life has to offer. In the decade since its inception, nearly five thousand people have settled in nearly one thousand municipalities. Far from policy perfection, many leave after their twenty-four-month term is up, unable to make lasting rural lives for themselves. Each earns a stipend, admittedly a meager sum. What many find is that theirs is a lonely choice. Rural communities provide little welcome for young people who seek other young people, nightlife, courtship, and a sense of belonging. In a few instances, the presence of host organizations improves the likelihood for Squad members to remain.

Spend even a little time in Japan and you will soon notice that migration patterns are top of mind. In 2019 the government relaxed national borders to accommodate guest workers for agriculture and the service industry. This minor shift is a notable aberration from the past, and perhaps reflects how desperate farms and shops are for labor. In other words, the needs are so great that policy-makers feel comfortable veering from the worldview shaped by the imagination of islands: Think of "Japan for the Japanese," akin to Brexit in Great Britain.

As a result, nearly all of the discussion of migration in Japan concerns itself with domestic migration: between cities, and from rural to urban. Whereas on the surface it may appear as though the cultural homogeneity of the nation is preserved, lean in closer to discover that regional differences matter greatly. As with elsewhere in the world, accents indicate geography and education. This means that villagers who strike out for life in the big city arrive carrying considerable baggage, most notably the stigma of small towns. Since Japan allows for almost no foreign migration into Japan, demographic discourse almost always concerns itself with how best to shuffle and reshuffle the deck of the existing population.

Much like the American civilian corps, AmeriCorps, the Japanese Squad program is imperfect. It is underfunded, and it is a policy success orchestrated by the opposition parties when they had attained power. These two elements alone should be enough to sink a modest program. However, it lives. Survival is itself an indicator of success. The Squad program survives because it addresses concerns that remain top of mind for most Japanese: life for rural areas and employment for the young. The Squad program elevates rural areas as possessing attributes valuable for the entire population. It places young, educated people in these areas to learn from the elderly so that traditional knowledge is not lost. It is also assumed that subsidized employment for the young reverses some of the brain drain that devastates rural areas.

Undoubtedly, the arrival of Squad members can be disruptive to a small rural village that is set in its ways. Squad members can be idealistic, arrogant, woefully unable to perform demanding physical tasks, or incapable of listening to elders. All of these may be true. However, the program represents a novel approach to repopulate areas in need of young people. Since there is national consensus about the urgency of how to improve domestic migration patterns, it seems wise to frame and measure the Squad's efficacy as a program that achieves multiple goals in order to attract additional resources. With additional resources, organizers can better prepare the communities for newcomers, prepare the newcomers to be cognizant of their roles, and invest in Squad members who wish to remain after their service is complete.

Take the J-Turn and Find Kuni

Road signs are used as a metaphor to interpret demographic trends in Japan: take the I-turn, the U-turn, and the J-turn. Largely focused on places in need of people, economic activity, and ideas, these signs are useful to determine precisely who will drive change.

The I-Turn: Those Who Visit a New Place and Stay

This turn describes individuals who, figuratively, put their finger on the map, punctuating it like the dot on a lowercase *i* in the Latin alphabet. While newcomers often introduce new ideas, their presence can be counterproductive. Unfamiliar with how to navigate local personalities or respect tradition, they stumble about like bulls in a china shop. And while this is an obvious point, it matters. Those who suddenly roll into town may soon roll out. Their mobility undermines the confidence people have in them to stick around and take responsibility for their actions. Squad members are I-turners.

The U-Turn: A Story of Regret

For individuals who grew up in isolated rural places with few opportunities, the universal plan is either stay and rot or escape to the city and find opportunity. Departures in Japan's post–World War II period are described vividly in Alex Kerr's book *Lost Japan:*

> When a family decided to leave their house for the big city, they would take practically nothing with them. What good were straw raincoats, bamboo baskets, and utensils in Osaka? Everything that had been a feature of life for a thousand years had been made irrelevant overnight. On entering one of these houses, it looked as though the residents had simply disappeared.[13]

While the dramatic abandonment of homes is no longer the norm, for those who leave today it is difficult to return. Yes, they return for national holidays and funerals, but to return permanently is difficult to explain to friends and family: "Why are you back? Could you not make it in the big and competitive city?"

The observation I heard from many working in community development is that the U-turn is a sign of failure. Return home, abandon the dreams you once held, and pick up where you left off. Continue the family work: rice farming, fishing, forestry, you name it. As a result, the U-turn is rarely a source for bringing back big ideas from the big city. Instead, accept the small town, small dreams, and the existing power structure.

The J-Turn: Almost Like Returning Home

As with the U-turn, the J-turn describes someone who escapes to the big city and returns. But the return is not the return home. After all, that would be too humiliating. Instead of returning home with big-city ideas, it is preferable to settle somewhere that is like home. It might be somewhere near home, but not precisely where you were raised. You avoid the whispers from family members and friends: "Why are you back, and how did you fail?" The return may also be somewhere similar to home, but on the other side of the country. This way, your knowledge of small places is a plus, but it is not accompanied with the baggage of coming home.

Why is this called a J-turn? It looks like the letter *j*. Moreover, the placement of the letter *j* reflects the return destination's location. It is not an *i* and it is not a *u*. Since *j* is located between those two letters in the alphabet, it has come to be known as a J-turn.

In places that are isolated and stuck in a rut, it is difficult to introduce new ideas. Local leaders are more concerned with preserving the status quo than entertaining anything new, frustrating those who seek change. Locals are discouraged from introducing new ideas themselves, too intimidated by the established order to challenge business as usual. In order for new people and new ideas to find hospitable homes in these dying places, outsiders provide navigation.

Sekihara's J-Turn

Tsuyoshi Sekihara grew up in rural Niigata Prefecture. Overlooking the Sea of Japan, Niigata is known for its snowy winters, thick forests, and rice grown in pristine mountain waters, destined for prized sake. It is also known as the prefecture hit hardest by Japan's population decline. In 2017, Alana Semuels reported in *The Atlantic* that Niigata "is expected to lose 40 percent of its women aged twenty to forty by 2040," a finding based on former Iwate Prefecture governor Hiroya Masuda's controversial 2014 book, *Local Extinctions.*[14]

Like many of his generation, Sekihara left rural Japan in the 1980s for opportunities in the city. He pursued an interior commercial design career in Tokyo. After some success, he experienced a midlife professional crisis and a marital divorce. He returned to rural Niigata. However, he did not return home. Instead, he found himself taking a J-turn.

Sekihara describes how ashamed he felt about his failures, and how he could not imagine relegating himself to the place and rules of his youth. Had someone suggested that his chance visit to Nakanomata, in the nearby municipality of Joetsu, a city and surrounding region of 200,000 people, would reshape the rest of his life, he would have shrugged with disbelief. Twenty minutes from the coast and nestled in the foothills and forests, his new village of Nakanomata is one of twenty-five or so in the immediate area. Sekihara describes the incidental nature of his visit. He was not searching for a new home. Rather, he was simply searching.

For those who are beginning careers, it may be difficult to imagine the kind of mid-career soul-searching Sekihara was undertaking. In truth, it is astounding that these ill-planned twists and turns do not occur more often. Sekihara's visit to rural Joetsu changed him, and as a result, set the rural hamlets on a very different trajectory once he began to make his presence felt.

Viewing rural Japan with fresh eyes, he began to recognize that natural beauty is not really so natural. Rather, it is manicured. Forest management clears out the undergrowth that otherwise would serve as fuel for forest fires, dead trees, and invasive species. Who is meant to manage forests today?

For centuries, the forests have provided locals with livelihoods: wood for fuel, animals for hunting, mushrooms and herbs for foraging. Long ago, it was the domain of the community to share in the responsibility to care for the commons. In the early twentieth century, local village governments began to assume greater professional management of place. Even after the great migrations of the mid-twentieth century, government budgets were able to subsidize this increased and increasing role for government management of community lands. Since the money was there, and able local populations were shrinking, it simply made sense for the government to assume this responsibility. This arrangement worked for a few decades. However, according to Sekihara, by the turn of the century, you could begin to notice signs of neglect if you looked closely enough: "road signs, once clean and shiny, had become rusted." His memory of beautiful rural Niigata contrasts sharply with the reality he experienced during his daily walks in the forest.

This is the spark. If the central government no longer sees the value in managing lands that once were critical for local survival, then whose responsibility is it now? Are there enough residents remaining to handle the task? To care enough to do something about it? Or is it time to adjust expectations and acknowledge that, despite the nostalgia for old ways, human civilization will abandon these parts? The forces of capital and scale would argue that places like rural Joetsu are moribund.

Were Sekihara's initial actions intended to spark a political campaign? Or was he personally compelled to care for neglected landscape and relationships? If a civic infrastructure once managed the forest and irrigation systems in rural Joetsu, why not revive it? The seniors remain. They have few years and options left. Meanwhile, give the children, who ponder escape, new opportunities to deploy their youthful energy, and who knows what is possible. Might they too share in Sekihara's desire to care about surroundings?

Akin to the Greek myth of Sisyphus, whom Sekihara cites in his poem "Kuni Manifesto," his actions may have appeared laborious and futile. At the beginning, only a small number of fellow inhabitants joined in. However, surprisingly quickly, he began to gain both respect and followers. He cultivated a culture of autonomy and mutual aid among neighbors. This triggered memories of pride for the beauty of place. His strange familiarity to neighbors—someone like them, but slightly different, having grown up in another part of Niigata and having pursued an urban career—allowed them to accept his certainty. While he may have had doubts, he transmitted self-confidence. Young people accepted the authority of his age and welcomed his strange outsider ways. This is the power of a J-turn. For two decades, Sekihara has reinvented the economy of Kamiechigo, and with it the social contract between its residents, between rural hamlets and municipal government, and with the repeat-visiting newcomers and fans from all over Japan. Sekihara's actions demonstrate the corny concept that individual leadership not only matters, it is many times only the primary element that matters.

Additionally, his experiment in local autonomy strikes a tune that many people are humming these days. The refrain is stuck on a pattern. It suggests that the gamble no longer works for me. Which gamble is this? It is the one that insists that big growth-driven institutions will carry us forward. It no longer seems worth it. To gauge just how out of kilter these relations are, we tend to turn to the throng of people who are likely to benefit the most from these bets: those in the big cities and employed by large corporations. On one hand, it makes sense to go with where the majority of people are to determine if things are working. On the other, if we instead turn to those who are likely to benefit the least—those left behind in rural communities too isolated and too small to matter—we may be better able to understand what is coming down the pike for us all.

Since Sekihara and I first met in 2017, we have had several intense opportunities to meet, exchange ideas, and traverse the divides of distance, culture, and language. Among the memorable observations he has effectively communicated to me, usually with the help of interpreters, one continues to haunt me. Our conversations often turned to road signs, and on one occasion in New York City, he approached a street sign. We were walking furiously between meetings in Manhattan. He motioned to me to come and take a look at the back of the sign. I believe it was "No Entry." He pointed to where rust was collecting at the place where the sign attaches to the pole. "You see, the rust is here."

With a certain measure of melodrama, Sekihara insists that we look for the signs that appear long before the shifts in public opinion or policy changes. In rural Joetsu, the robust infrastructure constructed by the state during times of explosive growth—by this, imagine bridges, clinics, new schools, roads, train tracks, and so on—today are rusted, forgotten. So too are the promises made. Sekihara's experiments in grassroots democracy began after the infrastructure had begun to decay. Today, he insists that we look for the rust now. Get used to it or remove it. Begin to reclaim some degree of autonomy as protagonists in the places where you live.

Kuni also addresses the question of place itself. Are we committed to a place? If so, who defines the boundaries of these places? Sekihara's quest for rural autonomy could draw a thick line around some communities and not others. Instead, with a hyperattentiveness to the scale of place (discussed in greater detail in chapter 6), he rejects the isolationist tendencies one would assume as central to any social movement driven by dreams of autonomy. Instead, kuni removes the barriers of community and opens it to all and any who wish to contribute to the region's future. More than making an elaborate plea for voluntourism,[15] Sekihara invites people to embrace place-polygamy, a concept introduced by German sociologist Ulrich Beck. Remain loyal to and engaged with more than one geographic location. In the age of short-term rentals and second homes, this invitation may sound elitist. However, consider the disruptions of the COVID-19 pandemic on work and geography. Many workers are returning home, where the sense of belonging may have felt like claustrophobia. During times of uncertainty, it feels welcome. Or consider the disruptions to child care in city centers. Head home for the wider support network of grandparents and siblings. Faded assets gain new color and appeal. These moves need not be the U-turn described above. Some are moving to smaller places that are like home, but not home itself. They are the pandemic's J-turn. In search of serenity, safety, and scale, households are rethinking everything, including location and work.

6

APPROPRIATE SIZE: When the Population Is Too Small or Too Large

Tsuyoshi Sekihara

After my experiments with the small regional forestry industry, I moved on to establish a Regional Management Organization. Before discussing the specific structure and function of an RMO, I want to share the fundamental reason why we need a kuni, which is built through the activities of an RMO. As I mentioned before, this is about scale, the issue of appropriate size. In other words, if the place is no longer an appropriate size and is excessively big or too small, it causes problems. Then, what is an appropriate size in terms of population? Based on my experience running an RMO for nearly twenty years, an appropriate size would be a population of more than five hundred to about two thousand. These numbers are not strict, and they could fluctuate depending on each specific situation. However, it is never less than one hundred or more than five thousand. Population is not just a number, and there is magic in it.

Magic in Population

Totally independent of religion, ideology, or social institutions, the population can change the social structure depending on its number. A society of a hundred people or ten thousand people form a society that reflects the population size. Most people would say that is common sense, but it's not as simple as that. The magic of the population is more powerful than we think.

Let me give you an example of an overly small community. Among many of the villages that I know, once the population shrinks beyond a critical point, I noticed the occurrence of a similar phenomenon. The power structure is fixed by hierarchy based on age. As a result, we see the emergence of a small dictatorship, in which dissent, counterarguments, and counterproposals are silenced. Rather than helping each other, people keep each other under surveillance. Communality is more important than ethics, and at times the people become complicit by working together on something unethical. The crime they commit is close to the "banality of evil" described by Hannah Arendt. Everyone is aware of the small crime committed by the small dictator but stays silent.

I am sure there is a society somewhere with a "good dictator," but as long as power stays in one place—with the dictator—society can head into darkness anytime. In a society like this, no one has the power to correct the dictator's behavior. They are no longer able to change themselves. The only way to remedy the ills of excessive depopulation is to increase the population. In this case, it means a number of villages need to come together to form a larger community. Once the population is larger, a small dictator from a village of fifty becomes just another member of the village. When there are more participants in the power game, autocracy disappears. This is not a fantasy.

I have actually met some dictators. They showed their true colors when they said they were going to grant us permission for RMO activities. They spoke as if they were the guardians of the village. By indicating they could grant permission, I could tell that they thought they owned the village. Legally, they only owned their houses and fields, but they felt they owned the village. Other people in the village didn't challenge them. This type of everyday dictator thinks that the RMO will cater to them. They think that they have the right to benefit from it. As self-proclaimed village gatekeepers, they demonstrated their deep-rooted inferiority complex toward the city. Meanwhile, the mass media eggs on this dynamic by portraying the villages with condescension. City people look down on rural people. Eventually, I came to recognize that when small dictators were granting permission to urbanites or the RMO, they were taking revenge on the city. I became a proxy for the city. One way or another, this distorted sense of ownership and permission breeds tragedy.

In one village, a woman who was subjected to this sense of ownership and permission was sexually assaulted by a small dictator. She came from the city with the dream of living with good villagers, but, in fact, the dictator acted as if he owned her and that she had given him permission to prey on her. The normally good people stayed quiet and became accomplices. There is magic in the population. When the community is extremely small, even good people become passively evil.

If the villages were called upon to join a kuni, would they abandon the dictator for the sake of their future, or would they stay with the dictator? It's up to them to decide, but I neither support nor bemoan the extinction of villages that are unable to break away from small dictatorships. The villagers are facing challenges. Do they have the will to live in the future, or do they sit by the window of a hospice at dusk? There is magic in the population, and you need to work that magic. In other words, the population size, one that is neither too big nor too small, is the magic of a kuni.

The New Wildness of Big Cities

Now, what happens to a society with an overgrown population? That is Tokyo, as well as New York. I lived in Tokyo for about sixteen years, and for a place with such a huge population, I hardly spoke to anyone. I think it was the same for others who belonged to small groups based on obvious relationships. In large populations, people tend to come together in small tribal groups. From a yoga group to a violent gang or criminal organization, we can say that they each turned themselves into tribes. These tribes are normally smaller than a depopulated village, and we can say that there are countless small villages in big cities. The reason that tribes are formed comes from difficulties in seeing larger numbers of people as human beings. Therefore, people end up forming a tribe with the number of people that they can perceive as human beings.

For example, strangers on a crowded commuter train, with their bodies pressed close together, are indeed people, but they blend into the background. In fact, there are so many people in the train car that you need to shut them out to stay sane. You prefer to expend energy to shut them out and make them blend into the background rather than in seeing them as human beings. However, no matter how much you shut them out, you are unable to eliminate the pressure of human existence. I call this the "pressure of presence." One good way to understand the pressure of presence is to imagine frog eggs. There are black spots on the clear gel-like balls. They are densely packed and pressed against each other. Imagine a crowded train full of frog eggs. The black dots are you and me. The transparent gel is the presence of others that you can't avoid feeling. No matter how much you try to escape from it, you can't shut out the presence. Therefore, you forcefully block things out.

Nowadays, I am rarely in Tokyo. When I am, however, probably like you, I ride the subway. It is common to see every single rider fiddling with their smartphones. Are they transfixed on their devices to escape from their surroundings? They do not look as though they are enjoying themselves. I grow somewhat self-conscious about my interest in their behavior, as I get lost in trying to understand what is going on. I have come to believe they feel the "pressure of presence." Are they trying to escape? Or are they a single neuron, impatiently pumping out tiny outputs for the giant brain of the system? They happily pay a small fee to continue working as neurons. Is it that the dopamine secreted by the giant brain is so strong that no one can stop it? This is where my mind goes on the subway: awry. While their eyes are fixed on the screens, mine are on them. In my head, I hear the lyrics to Simon and Garfunkel's "The Sounds of Silence." What do they hear?

If everyone in the city shuts others out, what happens to their mental state? I imagine that it's not positive. In a place with a large population, the magic of scale also comes into effect. However, I fear that it is a dark magic.

Prophecy of the Beast

This is my prophecy about the beast of the city. Though quite grim, do not take it too seriously. Instead, read it like a dystopian fantasy novel. In my search for the right-sized community, I cannot help but consider what happens when a place is too large. What happens when the population density exceeds a certain critical point? I anticipate that the ethical social balance deteriorates rapidly. The city transforms into a new wildness: living quarters in conflict, the prevalence of a law of the jungle, a myriad of buildings with countless invisible places, a cluster of stores that look like carnivorous plants, and the emergence of numerous urban tribes. Out of this new wildness, we see the birth of a new beast. The beast will start attacking others for no apparent reason. For the beast, we are just part of the background and no longer human beings.

People are controlled by vigilance and tension because of their fear of the beast. Punishment for beasts that are caught will become harsher, and a Code of Hammurabi for the twenty-first century will emerge. People who are afraid of the beast will insist on punishing the beast as a collective vendetta. People who live in the new wildness will shut people out. As a result, each individual becomes isolated. People will start to pour money into and work to maintain their isolation. Industries that help shut others out grow rapidly. Their stock prices rise. These industries provide surveillance equipment, security, robotics, and virtual reality. Broadly speaking, these industries help people remain wary of others and live without meeting anyone.

The bigger they get, the media in turn sensationalizes the anxiety caused by the beast and drives people to expensive shelters. Those who can live in shelters are the haves. They show off their wealth to set themselves apart from the have-nots. Ironically, the ones who live in these shelters brand their foreheads and make themselves the prey. What they try to shut out is not just the beasts but everything from bacteria to viruses, pollen, dust mites, house dusts, spores, humidity, oxygen levels, even smell. Living in shelters is to live in fear and to feel endless stress. When the stress level reaches a critical point, the residents in these shelters begin to accuse others of being beasts, even if they are not, and attack them. Just as in the Salem witch trials, the residents become beasts.

The state or democracy is not going to disappear, but the people will eventually break down into different groups. Increasingly, we have difficulty understanding one another. Division does not yield equality. On the contrary, the will of the upper class in the urban areas is reflected in public policy. The will of the lower class is not taken into consideration. Regional cities shrink further. Either they go extinct or become the provinces of some larger city.

The elderly's right to live starts to diminish year after year, despite progress in a health care system, of which only the rich now benefit. It becomes legal to prescribe medicine for euthanasia for those over the age of seventy diagnosed with a terminal illness or the onset of irreversible dementia. A new beast emerges from the elderly. With the right to live greatly diminished, this new beast is seen as a burden to the young. Feeling that there is no choice but to accept their fate, the elderly beasts are in despair due to aging.

Some high-income highly educated couples conceive babies through artificial measures and manage to eliminate any predictable risks for the fetus. If any of them have issues with fertility, they purchase expensive sperm or eggs possessing DNA that has passed quality tests. If you pay more money, you can edit your genome completely. These people from the future live in a shelter in a high tower. And, as stated before, among the people who were born with near-perfect genes in sterile closed quarters, a new beast will emerge.

If that were not terrifying enough, my concern for those without resources keeps me agitated on subways, as I watch people playing with their phones. Will we see a rise in unplanned pregnancies among low-income couples? Or a massive surge in child neglect or infanticide? I fear for a rapid increase in the number of infants in need of protection from their abusive parents. If they end up in the care of the state, will it see these infants as possessing inferior genes and, therefore, not spend money to educate them? These children will live their lives robbed of any potential and future. Parents who abandoned their children to the care of the state are then free to indulge in even more reckless activities. A new beast will emerge among them.

In big cities, countless small tribes are formed in self-defense, and these tribes adopt rules that have nothing to do with ethics. Superficial friendships and intimidation are two sides of the same coin, and violence and subjugation become the norm. Tribes in the city have their own fangs to protect themselves. Some tribes carry guns, some use hacking capabilities, and others use weapons to attack or defend themselves. Internally, these tribes face dangerous rivalries, and play small power games that trigger collective defections, betrayals, and killings of kingpins on a daily basis. Out of these tribes, a new beast will eventually emerge.

And it all comes down to the endgame. People start to understand what equality ultimately means. Rich or poor, you have the same physical structure and are equal as a living organism. There is no superiority or inferiority. If you stab someone, that person could bleed to death. Therefore, the beasts of the city can destroy others by taking advantage of what it means to be equal. The beasts have distorted illusions of themselves as superhuman and think they can kill others. At this point, libertarians who were enjoying absolute liberalism can no longer leave their shelters. They realize that ultimate freedom is "I can kill you anytime I want."

The Third Option

I will leave the judgment to you, but I hope you understand that a very small community or an excessively large community is not comfortable for humans. The problem is that we are unable to find the right size. I am not negating the freedom of living in a small village or a big city. There are many people out there who like that. It's up to the individual. What I want to say is that the problem is that you can only choose from these two options, and there is no third option. In that sense, a kuni could be seen as a kind of a right-sized shelter.

The idea of a kuni is not based on some fantasies about returning to the countryside. The most important reason behind a kuni is how seriously we can deal with the magic of the population. The real-life sense of crisis in modern-day Japan necessitates a kuni. However, the government can only understand rural revitalization through the utopian theory of returning to the countryside.

7

THE ROAD TO KUNI BEGINS WITH POTATO DIGGING

Tsuyoshi Sekihara

Let's go back to the steps that led to the establishment of my Regional Management Organization. Problems with regional forestry prompted us to establish a local industry using local resources. That in turn led to creating a forest nonprofit organization. As a result of these activities, I noticed that the root of the problem was not an individual phenomenon but a decline of the community itself. In response to that, the nonprofit Kamiechigo Yamazato Fan Club was established in 2002. When the nonprofit was established, I had no idea that the organization would include regional management functions. In fact, we had not fully grasped the concept and functions of an RMO. There was nothing that I knew from the beginning. I dealt with the problem in front of me, and then I could see the next step. Then I repeated these steps.

To me, it was similar to digging potatoes. It may look reckless, but that was the only way I could carry out things that needed to be done, and I think it was the right way. If you try to run *after* thinking, you end up not running in the end. I believe that the pragmatic way is to think *while* running. When you are operating in reality, it means that you are dealing with a reality that is constantly changing, and you have to make adjustments accordingly.

If you are caught in rigid isms, you go nowhere. You have to stay neutral to be able to see what is happening on the ground. Digging potatoes is a lot of work. If you don't dig out each potato carefully, one after the other, you are unable to see the next potato. The role of the RMO became clearer only because we took time to dig out each potato. That's why the functions of the RMO are meaningful, as it took a lot to figure it out. If the functions were created on a table in a conference room based on general data and prevailing rural revitalization theory, the measures would have been vague and pointless, and it would have looked as if it could be applied anywhere, and at the same time not applicable anywhere.

What It Means to Do Everything

The Kamiechigo Yamazato Fan Club started to fulfill its function as a Regional Management Organization five years after it was established. The profile of the organization at that time included the following:

The organization was an incorporated NPO with three hundred members (40 percent from the local region, 30 percent from the surrounding cities, 30 percent from the Tokyo area). There was a total of thirteen board members, eight full-time staff, fifteen part-time staff, and an annual budget of forty million yen. Half of the income came from projects outsourced by the city, such as managing the forest park and giving classes on environmental education. The rest came from industries making the most of local resources such as people, goods, and events. We were able to meet 70 percent of the twelve functions as an RMO.

Here are the twelve RMO functions that are necessary to manage a kuni:

1. Protect the livelihood of the residents: provide small-scale support mainly to the elders.
2. Maintain the culture: pass down, maintain, research, and archive folklore culture.
3. Small-scale welfare: provide necessary support such as exercise classes, opportunities to get together, and food to maintain the well-being of the residents.
4. Implement small-scale public transportation: operate compact shuttle buses that can take residents to the last stop on the private bus line.
5. Educate students about the region: teach classes and share formative experiences about living in the region so the children from the region will eventually want to return to the area.
6. Environmental protection: raise awareness on protecting the mountains, forests, rivers, seashores, animals, and plants.
7. Local industry using local resources: create industries using people, goods, and events in the region.
8. Publicly commissioned projects: manage projects for the local area commissioned by the local government.
9. Develop programs to create repeat visitors: the repeat visitors come from the city.
10. Become the main point of contact.
11. Undertake comprehensive administrative functions: carry out all organizational administrative work as well

as administrative work that is cumbersome for the local community.

12. Nurture human resources: educate young people, who will in turn manage the kuni.

These are the RMO functions we developed when digging potatoes. It may look surprisingly obvious after the lengthy explanation. However, more than each function, what is important is that the RMO is responsible for all functions. The local government tried to impose different organizations that would work on one function. They thought that they could just copy their own organizational structure. That resulted in a series of useless regional revitalization policies. These functions should not have been conducted separately. The heart of the RMO's functions is to encompass all functions. That was the most important insight. If you want to engage in regional management activities, the reality is that you have to do everything. There is no budget to create twelve organizations based on the twelve functions. Even if you do everything to scrape together a budget, it would be enough only to keep one organization going.

Two Types of RMO Work

The twelve functions of the RMO could be divided into two categories of work. One is to implement the local industry using local resources. The other is to undertake the small-scale public works. There are two goals for local industries. The first is to integrate local resources, which are residents, goods, and events, and to create a local industry using local resources. The other is to secure the operating costs of the RMO. The other work is to undertake small-scale public works, which the RMO does on a volunteer basis. The RMO manages the right-sized community and takes care of its small-scale public works. That means maintaining the health of the elders, educating the children about the local region, preserving the culture and folklore, protecting nature, and managing small-scale shared infrastructure. On the other hand, local government takes care of the large-scale public works.

As for the ratio of local industry to small-scale public works, it should be 7 to 3. Seventy percent of the work should focus on local industry, and the rest on public works. There is no statistical data for this ratio. It's just a matter of figuring out the ratio to maintain both work with income (local industry) and work without income (public works). I don't think this ratio is random. Any regional organization, as long as it is an RMO, would need to divide its work close to this ratio.

However, in reality, the balance is broken. The regional organization usually falls into chronic shortage of funds by working on small-scale public works, which does not generate income. But it is not an RMO if none of the public work is implemented. The reality is that the RMO struggles with this dilemma.

Small-scale public works and large-scale public works are inherently complementary. Since the local government focuses on big public works, the RMO can focus on small-scale public works. Also, organizations that implement big and small public works are different. They need each other. If the local government is aware of this, it will actively help out with small public works. In fact, the key for the local government to avoid bankruptcy lies in small-scale public works being done by an RMO. It reduces the costs to the local government significantly. Let me use elder care projects by my RMO as an example to explain this.

The region I live in is known for heavy snow in the winter, and the staff help with plowing snow. (There were incidents of elders dying while plowing the snow by themselves.)

There are no stores in the village, and the RMO staff take over the grocery shopping during the winter.

There is a lunch gathering every week during the winter. It's an opportunity to provide a balanced diet as well as a place to chat with friends and prevent depression due to living alone.

We organized a tea salon that took place more than 190 times a year. This activity was outsourced by the local government. People in the village come together to drink tea, talk, and laugh. We provided our staff with training for dementia prevention, and we taught residents how to stay healthy. We measured their blood pressure routinely, and the RMO staff kept track of everyone's situation.

We also provided residents with help on how to use the computer, printed photos from digital cameras, helped them fill out complex administrative documents, connected newly purchased DVD players, gave instructions on how to use a smartphone, and followed up on anything that the elders don't know how to handle.

The RMO office is always open. The office is situated in the region with full-time staff. It's a salon that is always open, and elders stop by regularly on their way to somewhere else. Speaking with the young staff offers a different type of excitement and conversation compared to conversing with other elders.

A Person Is Not a Number

That's all there is to it. At the heart of small-scale welfare services is calling a person by their name, learning about the person's life, family, friends, and occupations, knowing what the person is good at and what the person is worried about as well as the person's hopes. No one wants to be treated like a number that is given out at a health checkup. They want to speak as themselves.

Large-scale welfare services can't meet the needs and wishes of elders. Only an RMO, which works inclusively in a right-sized community, can deal with this. And the majority of these activities are not costly. The strength of the RMO is the thoughtfulness of its work thanks to the right size. Large-scale welfare needs to be thorough, but to be effective it is necessary to create numerous small-scale welfare providers. Both are important, and one or the other is not sufficient.

Elders as a Giant Resource

Are elders, who require a variety of services, a burden to society? In reality, it's the opposite. When I look back on my long experience of managing the RMO, it is clear that elders were the heart of kuni's human resources. We learned all the specifics of how to live on the land from them. There was so much that only they knew and very little that they didn't know. We learned all the survival skills—unwritten wisdom—from them, sometimes with words, sometimes with unspoken actions. It would not be an exaggeration to say the passing of one elder in the village meant the disappearance of a small library. In fast-changing times, we don't have the luxury to think about things that should never be lost. Once these things are lost, they will never return.

There are countless examples of how elders can function as the heart of local resources: how to make rice and vegetables, how to organize folkloric events, how to make and use folk utensils, how to create a waterway, how to fix a house, how to make a waterwheel, how to cook, how to preserve food, how to fish, how to hunt wild boar, how to look at the terrain, how to predict the weather, how to maintain relationships, and how to maintain vitality for life. They were teachers in every respect. The elders showed the power of the land to the young people, who could no longer perceive that power. They demonstrated ways to live in harmony with the power of the land through practice.

When we feel that our lives can depend on a particular piece of land, it provides a deep sense of security.

In this way, the elders are an invaluable resource for almost all of the twelve functions of the RMO. For example, the RMO provides environmental education to elementary and middle school students. The teachers are all elders from the region. During the classes, there is no abstract environmental education, and the elders focus on how to teach traditional skills for survival.

To learn these skills, the classes are organized based on the meanings given to each day of the week. In Japanese, all the days of the week have significance. To describe Monday in Japanese, we use the character for "moon." For Tuesday, the character is "fire." Wednesday is "water." Thursday is "wood." Friday is "metal." Saturday is "earth," and Sunday is "sun." The week starts with the moon and ends with the sun. Fire, water, wood, metal, and earth are in between. This was established in China around 1000 BCE, and it represents all the elements that make up the world.

The children are taught that all the elements represent skills for survival. The sun and the moon are the basis, and there are skills for survival using fire, water, wood, metal (as a cutting instrument), and earth. The elders know all the skills that make up the week. When providing care to the elders, it is important to remember that they are so skillful and there are roles that they can play. Being needed is instrumental to the elders' will to stay healthy.

Imagine a scene like this. One day, there are children at the salon for the elders. They are learning about the culture and history of the region. Or they are learning how to build a kominka, the building where the salon is held. They learn how to build fire for the hearth, and visitors from the cities learn how to grow vegetables, preserve food, and cook traditional dishes. The elders, the children, and the visitors from the cities all have lunch together. In the afternoon, they are taught how to make sandals out of straw. They are also taught that used sandals are not tossed as garbage but become fertilizer on the side of the field as they decompose. There are so many elements in this one scene, and by connecting the elders to the outside world, we can make this all happen at once.

Equilibrium, Circularity, Spiral: A Rationale for Existence

Elders not only have a rationale for production but also a rationale for survival. Three forms characterize the rationale for survival: equilibrium, circularity, and spiral. What I mean by equilibrium is to understand the amount of food that you need to survive and how to balance the amount of food you need to what you produce. You may think this is common sense, but if you ask someone who lives in the city how much food they need for themselves and their family to survive for a year, they usually don't know the answer. Elders also know that repeating the cycle of production and consumption, and maintaining the equilibrium, creates circularity. The circulation of water, crops, seasons, and the gods of farming all exist in this circularity. People who live this way create a beautiful spiral of life that neither increases nor decreases. They don't know cumbersome words such as *rationale for survival,* but they are living it.

Seven Hundred Million Yen for Just One Village

Local government is unable to provide thoughtful, carefully crafted activities for the elders. The RMO was able to do it since it was of the right size. In fact, the services provided for elders had a huge impact. I will give you an example. There was a village where the RMO was providing services in the region. The village population was sixty. In this village, the elders on average reached the stage of requiring elder care five years later than their counterparts in urban areas. That means the local government was saving three hundred years' worth of administrative costs for the care of the elderly (60 people × 5 years = 300 years). In reality, that is a cost that is paid out in urban areas. How much money are we talking about based on the cost of elder care in Japan? In fact, it's seven hundred million yen. One village was able to save seven hundred million yen. The RMO is working with twenty-five villages in the area. I would not say it's the same for all the other villages, but you can imagine the enormous amount of money saved.

The local government keeps track of income and expenditures, but they don't keep track of the amount of money they *didn't* have to pay. Therefore, they don't notice the enormous amount that they saved. This is not something to take lightly. In the near future, local governments will go bankrupt covering the cost of elder care.

Even without the analysis of a scholar, if you ask an elder for tips on how to stay healthy, they usually give the same answer: "I have work to do." You need to have an active role and you need people who rely on you. You have friends that you can talk to when you have a problem, you are able to spend a lot of time with young people and children, you eat fresh produce with no contaminants, you drink good water, there are seasonal events to attend, you laugh a lot, and you have opportunities to speak with outsiders who stimulate your curiosity. All of these things are supplied through the twelve functions of the RMO.

8

AN INDUSTRY WITH LIMITATIONS

Tsuyoshi Sekihara

Local industries using local resources are essential for the kuni. It is the financial resource for the Regional Management Organization and enables it to be self-reliant. A local industry using local resources is characterized by its small quantity and a variety of other items. The level of production is not big enough for a large number of unspecified customers. If one particular product sells very well, but the raw materials can no longer be supplied locally and are imported from another region to meet the demand, it does not fit the RMO's goal of establishing a local industry. The RMO should export local products to the cities, but it should never import the raw materials. The RMO must establish a local industry where most of its profits return to the local community. Purchasing and processing the raw materials must be done locally. Wages are paid locally, and part of the profit is used to operate the RMO. If the organization starts to sell products beyond their supply capacity just because they sell well, it will inevitably distort the process.

A Tractor at Night

Let me give you an example. In a mountain village in Japan, a project was organized for elementary schoolchildren to experience rice cultivation. The children planted the rice in the spring and returned to reap it in the fall. The amount of work that they did was just enough to experience the process, and the teachers and parents were enthusiastic about it since it was easy. The project was featured in the media and became well-known. Once the project became popular and the local community ran out of paddy fields for the project, one farmer caused an incident. Using the tractor at night, the farmer stirred up a paddy field where the children had planted rice during the day. Children who visited the following day planted rice again. This incident did not involve the local community and was an isolated case, but it is a prime example of what can happen when demand exceeds local supply. When they reached capacity, they should have turned down additional students. By doing so, it will actually result in building trust in the community.

The local industry managed by the regional organization has its limits, and it should stay that way. The RMO should be honest about disclosing the limits of the local industry. By letting consumers know its limits, it builds trust beyond the relationship of producers and consumers.

What the RMO wants is not a large number of unspecified customers, but name-based relationships with their customers, since they can only sell their products to a limited number of people. The goal is to build relationships initiated by the sale and purchase of products. In other words, it's not the RMO's work to engage in the sale of products that does not build relationships.

Once relationships start developing, it leads to repeated visits by the customers, and they begin to feel the sense of belonging to the land and the people. At this stage, we call them "repeat visitors." They are travelers and customers, but their motive to travel and purchase has changed. Their motive to eat safe and delicious rice has transformed into wanting to eat the rice from a specific local community. They know the mountain and the river in the community, the spring water that irrigates the rice fields, and the people who grow the rice. When the consumers' motive changes, it leads to closer relationships. If they decide to move to the community, they will lead a quality life. In other words, chances of encountering repeated issues after the move are low.

The local industry managed by the RMO needs to serve its role as the gateway for urbanites to enter the kuni. If the organization forgets that role and goes after immediate profit, it leads to operating a tractor at night. It is not easy to run an RMO, but when it comes to local industries, the dos and don'ts are quite obvious.

Self-Sufficiency as an Industrial Resource

Let me give you another example of an industry that goes beyond the power of the land. There was a mountain village in Japan that became famous for fruit trees. But the land was narrow and they ran out of space in no time. What did they do then? They started to recondition the land. They cut down the forest all the way to the top of the mountain and planted fruit trees. It worked for a while, but the neighboring village also started to plant fruit trees, and as the supply increased rapidly, the market price went down. At the same time, since trees had been cut down, the mountain could not retain water. This led to a decrease in irrigation water, which in turn had a serious impact on the rice fields. The community could no longer be self-sufficient, and eventually declined.

Regardless of whether it is the RMO's work, agriculture in a mountain village is completely different from industrial agriculture on the plains, and mountain villages should never ignore the power of the land and try to increase output beyond the means of the land. Agricultural produce from mountain villages with little arable land can only work in the market through added value. The core of their added value is self-sufficiency. If the village destroys its self-sufficiency, the mountain village has no choice but to disappear.

Let's look at local industries managed by RMOs and concrete examples of attempts to engage repeat visitors.

This happened in the "heart" village, mentioned in chapter 3. The regional organization received a request from the village. They said that several people were going to quit cultivating the terraced rice fields that year and, since they did not want the fields to go to ruin, they asked if the RMO could take over the rice cultivation.

The terraced rice fields had two major characteristics. The first was that they have been in existence for a thousand years. The village wanted to preserve their folkloric and cultural significance. The second was that the soil of the area was very soft, and if the fields were abandoned and the irrigation stopped, they would collapse. If the terraced fields collapsed, it would be extremely hard to restore them.

The regional organization decided to take over the rice cultivation, and it was decided that the RMO would open the paddy fields to anyone and start a terraced paddy field school. It was going to be difficult for the RMO staff to do their work and farm at the same time, and there were expectations that things might change if they decided to accept new types of farmers commuting from the city.

In a few years, new types of farmers from the city grew to about thirty groups. Almost all of them were families. It was not a seasonal experiential project, a rural Disneyland where they didn't show up until the fall harvest after planting the rice in spring. It was a full-fledged school where they could start growing their own rice the following year. That meant they would have to work on the fields almost every weekend. The size of the field that one family had to cultivate was small, but it was big enough to harvest enough rice to last a year for that family. Since each plot was small, the RMO could accept many families.

Unlike the gigantic mechanized farms on the plains, terraced rice fields with a thousand-year history have room for only very small machines. A piece of land that could be reaped in thirty minutes on level ground would take three days on the terraced fields. The skills that were needed were not knowledge of how to handle a machine, but knowledge of how to manually handle iron farming tools.

In the beginning, I thought that people would drop out, as they had to visit often and there was a lot of work. However, they seemed to think that the frequent visits were not a burden and were fun. For them, the manual technique taught by the elders in the village was magical, and they enjoyed learning the magic and being able to do it themselves. By repeating their visits, they learned that the terraced rice fields did not exist only to cultivate rice but existed as part of the folkloric culture. The elders in the village, who felt contentment teaching rice cultivation to outsiders, introduced them to the traditional folkloric culture still in existence.

These traditions played out primarily during the spring and fall festivals. Travelers were only allowed to view the festivals from a distance, but students from the terraced rice fields school were allowed to participate as part of the family. They helped prepare for the festivals, participated in them, and were allowed to enter deep into the shrine and sit close to the *kami* (deities). The kami of harvest comes down from the mountain in the spring and brings abundance, then returns to the mountain after the fall harvest.

The kami that we are referring to is a natural force, and people know that a natural force has two sides: nurturing and destructive. That is why it would be dangerous to let the kami stay in the village year-round. Once they are done with the harvest, they express their gratitude and ask the kami to return to the mountain. The kami also trace the circularity of the seasons. The repeat visitors from the city were surprised that they were allowed to participate in an occasion like that. It was something that they had only seen on television. The village festivals were an occasion for villagers to strengthen their community. Once the RMO was active in the community, the festivals started to function as a way to link insiders to the outsiders. It became a place to share their awe for the power of the land.

Selling Rice as an Insurance Policy

By opening up the terraced rice fields to the outside world, the project developed to the next phase. The regional organization started to sell the harvested rice. The RMO named the rice Uen no Kome. *Uen* means "there is a bond." When there is a bond with someone, you know the person by name. The Uen no Kome project was simply a way to have people purchase the rice.

The rice is not inexpensive; the price includes the operating cost of the RMO. And the rice is not a luxury brand—it is ordinary rice. Therefore, there are not too many buyers. The buyers are those who visited the village in the past, someone who had a connection, and repeat visitors with a sense of belonging to the land.

The RMO created many opportunities other than farming the terraced fields for urbanites visiting the village. The outsiders participated in various folkloric events, helped with trail maintenance in the forest park, renovated the kominka, made folk tools with straw, restored a charcoal hut, opened a small farm for sale, helped sell local produce at a stall, and opened a kominka café, among other things. The RMO created many opportunities for repeat visitors to participate in local activities. Many participants ended up becoming members of the RMO, and some of them became devoted repeat visitors. The Uen no Kome project was carried out for members and repeat visitors.

The project sold rice, but the goal was to demonstrate the self-sufficiency of the village and the rice that symbolized it. Self-sufficiency means you can survive without relying on others. Mountain villages are usually self-sufficient. The RMO tried to turn that ordinary yet potent force into value. If they can sell the power that mountain villages have, that means it is possible to create a kuni anywhere.

Self-sufficiency also offered help to urbanites feeling a sense of anxiety. It could function as an insurance policy. In other words, the added value the rice production provided urbanites was peace of mind. No matter what happens in the city, as long as they have a bond with the village, they can survive. The insurance is not about paying a premium every month and getting a payout when something happens. Self-sufficiency was the insurance product.

The most symbolic insurance function that the regional organization came up with was evacuation: urbanites could seek temporary refuge from the city in the rural area. The contract was to accept people who purchased the rice in the event there was a disaster, such as an earthquake, and they needed to evacuate from the city. This meant guaranteeing housing and food for the evacuees. In Japan, this is not idle talk—it could happen. Families who don't know anyone living outside the city could face tremendous difficulties in a postdisaster city.

In that sense, they have insurance. The RMO calculated how many people could stay for the long term in a facility that they owned or managed. That number capped the number of buyers for Uen no Kome.

Capping the number of people who could evacuate to the village was so much harder than figuring out the limits of the power of the land, but guaranteeing the insurance was far more important. When I did the calculation, I figured that we could accept one hundred people. I don't think that number is small. It's better to give a specific number of people, in our case a hundred, than to offer a thousand without being able to back it up. The RMO wants to secure a set amount of money to continue their work and is not interested in making infinite profit.

Bound Together, Actively or Passively

There were repeat visitors who went further and eventually decided to migrate to the village. There are not many examples, but I can explain the reasons people migrated to the village as a result of the RMO's activities. One migrant decided to move from Tokyo when he retired from his job and built a house by himself. He is an audio technician and listens to famous jazz recordings at home looking out a big bright window. He owns a house in Tokyo, but rarely goes there. In other words, the house in Tokyo became his second house. He didn't want to build a house in a mountain village to stay there occasionally. He wanted to live in a place surrounded by folkloric culture. When you decide to live in this kind of community, it means you will be given a role. He felt good about the human relations in the village when people worked together to clean the rivers, organize events at the shrine, and preserve the waterway, and he decided to migrate to live that way.

It's not so bad to be bound in this way. For introverted people, doing work together is to be bound by obligation unwillingly. For people who enjoy these types of activities, they are bound by obligation willingly. This eventually turns into a sense of belonging. If you visit a riverbank many times to cut weeds, it becomes your riverbank. In fact, everyone who worked there felt the same. It did not take a long time to notice that the riverbank was "ours." Repeat visitors said that they felt a sense of fulfillment when they engaged in physical labor to build infrastructure in the village. When everyone worked together, it did not matter if you were rich or poor, or had higher status or not. No one is an employer or an employee. Everyone had the same tool, was covered in mud, and participated in the evening feast. Everyone was equal when engaging in these types of labor. By sharing the experience, it created a sense of "us." This is not something that can easily be understood, no matter how much you explain the concept, because it is not something to be understood—it is meant to be experienced.

The Need for Consensus and Criteria

Most depopulated places say they want migration from the outside, but, in fact, they are missing something crucial. Within the community they had not created the criteria for accepting outsiders. The following are the criteria that my RMO developed:

"How much assimilation are you asking for?"

"What kind of programs exist for assimilation?"

"Who is going to implement the programs?"

"How do you control excessive demand for assimilation?"

In other words, how do you ensure their freedom in their daily lives? It is a given that basic human rights and freedom are protected by law, but everyday life in the village is controlled by local rules. The residents need to build a consensus on creating criteria about which rules are excessive, adequate, or discriminatory for outsiders. The criteria should be announced to the outsiders, but most communities put them in a black box. A kuni needs to make the criteria public.

The criteria could differ depending on the kuni. That is individuality. It is a problem if there are no criteria based on consensus, and when they are not made public. If the kuni makes the criteria clear, it opens up new possibilities. This means that, for example, the kuni has its own criteria for migrants. The criteria can include rules for the land to stay open but maintain its individuality, to maintain the balance between understanding other cultures while encouraging assimilation, to implement programs designed for a certain level of assimilation. If the kuni succeeds in making those criteria, it would not remain depopulated. Needless to say, an RMO with a comprehensive understanding of the village is essential to discussing, summarizing, building consensus, and disseminating the criteria.

9

THE FUTURE OF KUNI

Tsuyoshi Sekihara

The concept of kuni is not about rural revitalization. Even if the region was revitalized, if the cause of the decline is not eliminated, the region will inevitably decline again. Is kuni then about invigorating the region? If the region is in relatively good shape, it could be reinvigorated. However, rural areas have been in bad shape for a long time. Is kuni then about creation? Exactly. It is about creating something new. By fusing the old and the new, kuni is about creating something new. It may sound as if kuni exists everywhere, but it is nowhere to be found yet.

Coming Back to Your Senses

Most developed nations are aging. The number of elderly people is increasing, while the number of children is decreasing. The economy is stagnant and rural areas are shrinking. Only big cities are becoming hypertrophic, and the new wildness is expanding. In this reality, we neglect the power of the right-sized community. Things are gained by expansion and things are lost by expansion. The biggest thing that we have lost is the power of self-government. The sense of "we" has lessened in populations that have grown so large. Self-governance means governing ourselves, but if we lose the sense of "we," it is difficult to maintain self-governance. Under these circumstances, any sense of unity becomes detached from the land and is replaced by abstract concepts such as nation and race. People are more caught up with simple and incendiary slogans about an enemy or an ally, which are a mirage, than with their own lives. Ideology is a kind of fantasy. Therefore, an ideology that counters ideology is also a fantasy. The struggle between these fantasies becomes *The Never-ending Story.*

On the other hand, a right-sized community is a concrete place. It is a place where each individual is not separated from the land and can feel a sense of "we." In a place like that, people come to their senses. Once people are sane, they question the dubiousness of ideology, and words such as *nation, ethnicity,* and *race.* The state is not going to disappear, but there is no need to worship it excessively.

The Third Option

The problem is obvious: a near future in which people can only choose between depressed rural areas and gigantic cities. It is necessary to have the third option, the kuni. In a kuni, the power of the land is appreciated, and experiencing the abundance and hardship of the land brings people together like gravity.

However, the gravitational pull is weak. It tends to get dispersed in the presence of a strong power such as internal conflicts and tribal wars. When a strong power is in full force, it is hard to feel the weak force.

In other words, kuni is needed in developed nations that are aging. Aging means the deterioration of cells. Without stopping the aging of cells there is no way to stop the overall aging process. Kuni can function with the modern communications revolution and an advanced transportation infrastructure. The conditions needed for kuni to emerge are democracy, science and technology, transportation infrastructure, the communications revolution, a declining birthrate and aging population, a stagnation below the surface, the death of culture, and citizens dichotomized by ideology. The glory is in the past, and anxiety due to decline is hard to eliminate. It is today's aging developed nations.

The Importance of the String

Let us review the structure of a kuni. A kuni emerges in a right-sized community. Remember the magic in population? However, that is not the only thing that makes a kuni. A kuni is not a place to revitalize, but it is a place to create. In order to make it happen, the emergence of a Regional Management Organization is essential. (If an existing local organization is doing a good job, the community is not in decline.) RMOs implement twelve simple functions. The functions should not be divided, and it is important that the RMO implements all of them. The core of the RMO's function is integration and comprehensiveness. The act of connecting things will make that happen.

I often liken this to a necklace, and what is needed is not a jewel but a string that threads the jewels together. No matter how many jewels there are, you can't make a necklace without string. In other words, without string, things are not connected and don't work together. The power of integration brings things together and makes them work together. In the past, rural revitalization has been all about finding the jewels and ignoring strings.

Revival of the Five Senses

The work of these strings is brought about by the power of integrating the twelve functions of the RMO. The root of the power of integration comes from the five senses. If we are not in tune with the input from the world, our ability to integrate deteriorates. The deterioration of the five senses is not just in relation to nature. We use the five senses to comprehend other people's feelings. If our senses deteriorate, we can no longer understand other people's feelings. If our senses deteriorate, relationships become cumbersome, and a will for a community will not emerge. If each individual does not revive their five senses, a kuni cannot emerge.

A kuni is a highly self-sufficient place close to nature. Nature is also a place where you encounter death. There are no hand railings on cliff paths deep in the mountains. There is no enclosure in the sea where the tides meet. In these places, we mobilize our five senses and integrate all of them to survive. If we don't, we will die. In kuni, the five senses are revived and become stronger.

The five senses are about seeing, hearing, touching, smelling, and tasting, but once they are integrated, we are also able to sense the invisible. For example, when we are deep in a forest, we may suddenly feel that the forest is looking at us. This is because of our integrated five senses. We are in awe of the gaze from the forest and are reluctant to enter the forest further. That feeling helped preserve the environment in the past. It is not that humans preserved the forest because they thought they dominated it. They did not touch the forest because they were in awe of something that they could not dominate. When we look at a forest from above with Google Maps, there is nothing awe-inspiring about it. That is why people are able to go into the forest and destroy it through rampant development.

The ability to sense the invisible can be applied to our relationship with others. No matter how much we discuss community theory, if our sensibility to feel others has deteriorated, there is not so much we can do for the community.

The Image of Kuni

What shape do kuni form as they emerge through the power of integration and comprehensiveness? First of all, many kuni would encircle a small regional city, much like multiple moons surrounding the earth. The kuni, or multiple moons, are all equal and in equilibrium. The earth does not subjugate the moons and the moons do not exist on their own. The earth and the moons, the regional city and the surrounding kuni, are part of a single system. These independent systems can emerge all over the country. They are ordinary and not special. They are also complemented by repeat visitors from the big cities, but they are not subjugated by the big cities. The kuni systems are loosely connected, but each exists independently.

The image of kuni can be explained by a certain vision. In Buddhist doctrine there is a saying: "The small one includes the world, just as all of the world is included in a small one." This vision is explained in the story of Indra's net. To learn about humankind's good and bad deeds, the god Indra cast an infinite net over the heavens. The knots that tie the net together have a small beautiful bead. The surface of the bead is as smooth as a mirror, and each bead reflects off all the other beads on the net. One bead includes everything in it, and everything is included in one bead.

This way of being is the vision I have for kuni. I don't think it is a fantasy. This vision will no longer be a fantasy if we move away from wavering values, feel the assuring power of the land, understand the value of the right-sized community, and add innovations in modern technology such as communications and transportation. If a new structure like this does not emerge, I don't think it is possible to stop the inexplicable decline of developed nations.

Sisyphus of Our Time

We are entering the final phase.

Finally, it would be foolhardy to avoid a discussion about the deterioration of the RMO, its causes, leaders who can overcome the deterioration, and a human quality I call "Sisyphus." This is based on the ancient Greek myth of the king Sisyphus, who was condemned in the underworld to eternally push a boulder up a hill, only to have it roll back down when he finally reached the top. It represents at once determination and futility, but just as important, it can also be a metaphor for a cycle, and for the unending work of maintaining a kuni.

For example, a neighborhood association and an RMO are complementary. It's like the ebb and flow of a tide. The RMO will prevent the neighborhood association from narrowing its view, and the neighborhood association will prevent the RMO from deviating from the local character. These two entities should work like tidal dynamics, and neither one of them should become dominant. Furthermore, repeat visitors and an RMO would also have similar dynamics. The RMO will prevent the repeat visitors from ignoring the local character, and the repeat visitors will prevent the RMO from being buried under local character and narrowing its view.

One negative aspect of rural areas is strong pressure to assimilate. Just like gastric fluid that dissolves everything, staying in the same place will make you assimilate. Once you reach that stage, the RMO will fall into obediently following customs and inertia. I know of an RMO that started to have these symptoms around 2010. The telling sign was that the organization started to block the opinions of repeat visitors at member gatherings. They said, "You are an outsider and can stay here because we gave you permission. You don't have the right to voice your opinion on how to manage the RMO." This is "darkness in the village," which I discussed earlier. The RMO tried to change the darkness that existed, and it became an open space to survive, but then the RMO itself became the darkness.

This is a classic example of an RMO deteriorating. Arrogance that comes from doing something good for a rural area can lead to indulgent self-righteousness. Becoming self-righteous is not the scary part. Not being able to see the self-righteousness is the scary part. In a kuni, repeat visitors are essential to sounding a warning to the RMO.

While it may be exciting to explore how RMOs are established and managed and succeed, few talk about how RMOs decline. Most stories end with success. However, communities keep changing like a living organism. There is a distinct cycle. If there is success, it is natural that there is also decline.

It is important to know that this cycle exists. The cycle is not a closed circular movement where you eventually return to the starting point. The cycle is like a spiral with an upward force. In order to adjust to this cycle, the RMO needs to shed its skin like a snake. If you remain in the old skin, all you are waiting for is death. It is necessary to shed the old skin in response to the cycle. There is nothing to be afraid of. Even though they shed their skin, a snake is always a snake. The RMO should be afraid of the day when it can no longer shed its skin.

Who should prompt the RMO to shed its skin? Anyone who is complacent and has a certain degree of success would argue against it. Who would insist that a snake that no longer sheds its skin is not a snake? And what kind of leader is that? It is probably a leader who constantly asks, "Where is our community?" The leader does not condone putting process over the ultimate goal of creating community. It is a fundamental question for the kuni.

Community Intention and Will: The Intention Is to Live Together

In reality, the answer is not found in the community. Any existing community becomes distorted due to opposing forces like happiness and misery, wealth and poverty. From this, dark emotions can surface. This can happen in any community, and it makes me think that an ideal community can only exist through the will of an individual. But can it really exist? I am sure readers would question that. However, I am convinced that it exists exactly for that reason. I don't think an individual's will to make a community is an isolated delusion. One's intention for a community obviously includes other people. If I were told to prove the existence of an ideal community, I would say nothing like what exists in this world. However, if we were to consider the will of a person as something concrete, then a community can exist firmly in a person's will.

Community is built through intention and does not simply mean having a large number of people. If we are to question what a community is, we should first ask why humans want community. There is something that makes people yearn for community and because of that, a person will possess the determination to create one. I think this comes from fully understanding solitude.

Loneliness is an emotion that people start to feel more strongly as they get older. When you are young and nurtured by adults, you are united with the world around you. There is no such thing as solitude. As you get older, detachment begins. You get detached from the world—from your parents, your friends, and your lovers. The process of growing is a process of detaching. When you are detached from everything, you reach complete solitude. This is so important that you can consider it a second birth. No matter how difficult this process is, people are born anew as they embrace complete solitude. Once they reach that stage, people can fundamentally understand others and want to live with others.

Understanding complete solitude is closely related to understanding death. When people become aware of their impending death, they understand what tears people apart from each other. Therefore, the will to spend their lives with others emerges.

People want community and try to build it. They try various methods to build one, but the minute they start trying, distortion starts. When the individual will for coexistence is placed in the context of living with others, there are countless ways in which each individual will want to form the community. While pushing forward to coexistence, pushback will take place at the same time. To deal with antagonistic individual demands, the community will give in to simple solutions, which seem to be based on majority opinion and seemingly useful for everyone, but in fact are useless to everyone. This is the reason the community becomes distorted.

Still, when we form a community, it can only be found in the will of individuals. A community needs the will of an individual to live together, and that will is the community. Attempts to form a community will fail many times since the cycle never stops. And every time it fails, the snake needs to shed its skin.

Is this a tragedy similar to the myth of Sisyphus? Doesn't it demonstrate the triumph of man? The myth is a metaphor for an endless cycle of trying to create a community based on will, and you are allowed to try indefinitely. For people who are fulfilled by the act of trying, not having a deadline is not a curse but gospel. I call people who are aiming for a kuni the Sisyphus of our time.

A small kuni can only emerge by counting on a Sisyphus with a resilient spirit. As for me, I can surely hear the footsteps of these countless young Sisyphi. They could be in the midst of setbacks. I want them to know that they reach their potential to become a Sisyphus exactly because of those setbacks. The paths that each of these Sisyphi walk are probably modest, but there are no footprints on these paths.

10

CONFESSIONS OF A PLACE POLYGAMIST

Richard McCarthy

Tsuyoshi Sekihara's story sparks many ideas for me. One, Japan's rural crisis is further along than ours in North America. Why not glean insights from their desperate context? Two, he details the personal risks to act constructively in unpopular places. Beneath the bluster of certainty, individual initiative comes with doubt. This is something most of us hide. Three, Sekihara contends that there is a proportional relationship between the size of a community and democracy. The larger the community, the greater the freedom; however, the smaller the community, the greater the access to power. There must be a right size. Precisely where that right size is, between the loneliness of the megacity and the tyranny of the village, demands discussion. And four, place is a process. Sekihara succeeds because he recognizes the importance of flow between places.

It is challenging for small places to establish healthy working relations with large ones. It reminds me of the difficulty a tugboat experiences mooring to an ocean liner. The relationship is as uneven as the footing. Not only does Sekihara provide a strategy for the tugboat to moor itself, but he also discovers how to get the inhabitants on the ocean liner to long for life on the tugboat. Even better, he gets them aboard repeatedly. Considering all of the amenities aboard the larger vessel, this is no small feat to attract them to the lesser destination.

Pirates and the Two-Loop Theory

Keeping with nautical metaphors in order to dive deeper into this issue, consider pirates. My fascination with pirates begins in the Caribbean with the seeds of modern multicultural democracy. It also brings me to the English Channel in the 1960s. This is when pirate radio took on the stodgy British Broadcasting Corporation (BBC) and blanketed the airwaves with underground and African American music, largely ignored by the BBC. However, it is Queen Elizabeth I's pirates that are most relevant here.

In one of the seminal victories in asymmetrical warfare, an ad hoc navy defeated the "invincible armada" of Spain in 1588. King Philip II had sailed as many as 150 Spanish galleons to the English Channel in order to invade an emerging competitor, England. It is worth remembering that the Spanish galleons were revered and feared on the high seas. No ships were larger. It was also assumed that no ships were better armed than the ones that sailed beneath the Spanish flag. Imagine 150 galleons off your coast.

At this point in British history, Queen Elizabeth I had no formal navy. However, she had Sir Francis Drake and Lord Charles Howard. Together, they assembled a ragtag force of privateers and merchant and fishing fleets to take on the Spanish. As legend has it, the seas were stormy. Winds confused the Spanish and blew their ships northward. The English were armed to the teeth. Importantly, while their vessels were considerably smaller than the Spanish, this helped to make the English fleet more agile.

Upon learning about this naval battle in history class at school, I became fascinated with how a small scrappy navy defeated an empire that otherwise appeared invincible. I also became consumed with the thoughts that were going on in the Spanish sailors' minds. Did they look out into the rough waters, wondering if their fate would be better if aboard smaller ships?

As a result of the victory, the English navy came to be known as Elizabeth's Pirates. Ironically, roughly two hundred years later, British forces came to resemble the bloated Spanish military. In 1815, a ragtag military force, organized by pirate Jean Lafitte, defeated the English in New Orleans during the waning days of the War of 1812 between the British and the Americans. The British lined up in unison, proving to be easy shots for the American forces. With home-field advantage, the Americans utilized their familiarity with the bayous and marshes in order to defeat a larger and better equipped military force by deploying the kinds of guerilla techniques that helped to give the English navy its moniker, Elizabeth's Pirates, in 1588.

While the point may be obvious, Sekihara and I are not engaged in asymmetrical warfare. I seize upon the opportunities to build the unlikely relationships between the farmers who hail from small places and the consumers who inhabit large ones. Though Sekihara may refer to Tokyo and other megacities as beasts, he is not trying to take them down or lead a secessionist movement away from them. Rather, we are both intrigued with how the interactions that occur between differently sized communities mutually benefit the residents of both. Magic is found in the transactions. Moments to treasure, they unlock critical levers for transformative societal change. They provide unexpected and unexpectedly personal experiences for change.

Conventional thinking fixates on scalability as the ultimate goal to deliver returns on investment. This line of thinking may admire the same interactions. However, they would consider each as opportunities for scaling up. In other words, how can we help the little people and little projects join the world of the large? If they encounter a promising practice, like Sekihara's Rice Covenant, the logical next step would be to evaluate how such an innovation can expand vertically or horizontally. By contrast, we ask, what is lost when rushing to scale?

Proximity.

Some projects work precisely because they are intimate. They are built upon transactions that reward proximity. Sekihara has told me time and again that this is what motivates repeat visitors who travel from Tokyo to rural Joetsu to participate in community rituals like the harvest festivals. Uninterested in big events, if anything they run away from anything and everything that reminds them of the big and muscular places they call home. Instead, they yearn to be embraced by the warmth of a village.

Similarly, I see this same desire to join with others as a primary motivator for participation in alternative food distribution strategies. Farmers who organize their lives in order to market the fruits of their labor in town are nourished by the relationships they forge with the shoppers. Feelings are mutual. Shoppers also seek aberrations from their everyday lives. They are motivated to support specific people with whom they have grown familiar via repeat purchases. Choices become habits. Habits become new norms. Similar sentiments exist in other piecemeal strategies: community-supported agriculture, community gardens, pop-up restaurants, food trucks, and even the revival of door-to-door delivery. In each, participants attempt to give meaning to lives and livelihoods through direct contact with those who live elsewhere. Retail anthropologists may coldly remind us that these are natural tendencies in economies to accommodate post-Fordist consumption habits. This means that those with resources seek out superior products and experiences in order to separate themselves from those who still have to settle with the mundane and mass-produced clothing and food. I contend that this yearning "for something more" is not just for the wealthy.

In the United States, the pandemic has triggered the kinds of existential questions that make workers and families ask: "Is what we are doing with our lives really working?" Or consumers ask, "Why do I buy?" Could it be that the irregularly shaped strawberries on offer at the farmers market speak to people at an even deeper level than first imagined?

I think back to the first season at the Crescent City Farmers Market. A winter cold front had walloped the strawberry farmers north of New Orleans. During the days between markets, I received a call from a shopper I knew by name. She expressed her concern for one of the families whose berries she had come to rely on each and every week. I was a bit surprised and encouraged by her concern. I did not even know she knew their names, let alone to watch weather reports on the local news with the farmers in mind. Proximity, it seems, bears fruit.

Or consider Sekihara's Kamiechigo Yamazato Fan Club members in Tokyo. When I described to him this story of strawberry solidarity and how the phrase itself became a rallying cry for my staff to work harder to deliver the intangibles people crave, he recounted with his own recurring refrain, one that urban Fan Club members would ask him repeatedly: "Why do I feel so much more at home here than I do at home in the city?"

This brings me back to Queen Elizabeth's Pirates. How alike are the passengers on the ships of the Spanish armada to Sekihara's repeat visitors who travel for working weekends from the beast-like megacity of Tokyo to Joetsu? Or consider the consumers at urban farmers markets. They dream of rural livelihoods.

The strawberry solidarity example captures that ability to reside here and yet care about there, as if it is yours. However powerful it may be to express urban care for rural farmers, the woman who called demonstrated concern, but did not suggest plans to help on the farm.

Change is risk.

What about the farmers who have not taken the risk to direct some of their business "off the grid" and away from faceless conventional wholesale channels and toward direct marketing? Nearly everything about farming is risky. Just take into consideration the weather. Moreover, the longer a farmer holds onto products, the risk of losing value increases by the minute. In this regard, the farmers market vendors play a high-risk hand. Despite these risks, farmers are looking for a way off the grid, in search of greater control over their destinies.

On multiple occasions in Louisiana, local agriculture offices invited me to meet with farmers who had prodded civil servants to facilitate meetings with the organizers of these emerging markets: me. In general, these farmers were larger and better positioned to leverage conventional wholesale channels. They were also better able to garner the technical help from public agencies to succeed. Despite their advantages, they took great pains to recount their impressions of the success neighboring farmers were enjoying via direct marketing at farmers markets. They noticed the indicators of apparent success: the purchase of larger trucks and new equipment. However, what really got their attention was this: a multigenerational excitement for new opportunities, especially as newspapers began to publish stories about the new face of agriculture.

They pressed me for details. And while they may not have expressed it overtly, they alluded to an exhaustion for their current model, with language like "dark days ahead." In at least one instance, the farmer provided me with excruciatingly detailed accounts of how an uncle had lost his farm because he was overextended. Everyone nodded in agreement. Then they quizzed me: How does it work? How much volume do we move? Who decides what? And so on. Questions kept coming. At some point, I realized that I could not adequately answer the primary question on their minds: Is there room for me there beneath the tents and umbrellas in New Orleans? To be honest, at the time, I did not know. I, like Sekihara, had taken the leap of faith that an alternative path can be built. However, I did not possess sufficient evidence to mitigate all of the risk for all of the farmers. As sloppy as it may sound, I was operating with more art than science. I may have conducted research to build a case for new modes of commerce built on very old models of direct marketing; however, I did not have the resources to construct a watertight model with complete certainty.

My work, like Sekihara's, entails risk. Who is prepared to take these risks? The obvious answer is those who are ready. Maybe they are too old to care anymore. Or they are too young and careless to understand failure. Or even better, they have been waiting for us. The farmers I met on this one evening were not any of these. Instead, they were greatly burdened by debt they had accrued operating in the old model. We were venturing into a new one, and they were trying to assess on which ship they would sail.

This is where pirates meet the Berkana Institute's two-loops theory. Cofounded by visionary organizational development authors Deborah Frieze and Margaret (Meg) Wheatley, the Berkana Institute is named for the ancient Norse word for the birch tree. Its name reflects the institute's interest in the growth and rebirth in communities and in organizations. Always attentive to the edges of where change occurs, Berkana has cultivated a following across various sectors.

After Hurricane Katrina, at Market Umbrella in New Orleans, we were fortunate to begin a relationship with the W. K. Kellogg Foundation. One of the perks upon becoming a grantee is that you are invited to attend gatherings where you meet other grantees, funders, and leaders. In one of the meetings I missed, Deborah Frieze presented Berkana's two-loops model for nonlinear change in complex systems. While the model may apply to the growth and decay of individual organizations, it also applies to sectors—like mine, the food movement—and to civilization as a whole. Colleague Darlene Wolnik attended the meetings in my place. She returned from the meeting pumped. Wolnik acknowledged that "I have no idea whether anyone else grabbed what I did from the presentation, but it's just like what we've been saying about pirates."

Wolnik did her best to recount what she had heard:

> Most of us accept the mechanized view of change. As with machines, if a part no longer functions, simply replace it. With new parts, innovations are introduced and the machinery resumes. The problem with this worldview is that it fails to take into consideration that organization and sectors are like living systems. Systems have cycles. They die when they no longer meet people's needs. New ones emerge; however, the transition from one to the next can be dangerous. Just consider the transition from feudalism to capitalism. There is more to it than switching off the lights here and then on there. Many wars, famines, political struggles, and disruptions later, here we are! In fact, we spend far too little time discussing the difficult stages of replacing one system with another.

At this point, I urged her to go on. Living in a broken city, it was easy to imagine what collapse looks like. I did not need Jared Diamond to remind me of the fragility of systems that are unable to adjust to change, although he sure helped to remind me that this show has run before. Complex societies grow too complex to manage change, even if generations earlier could have.

Whereas Diamond provides historical context for collapse, Frieze and Wheatley provide a model to understand the unfolding events of growth and decay as they occur in real time. If change strikes terror into people's minds, try collapse. Not only are stakeholders deeply invested intellectually and financially in the old ways, so are we. While nobody likes change, we can manage and mitigate its most uncomfortable effects if we understand it.

Berkana's two loops refer to the downward cycle of decay occurring at a time of an emerging upward cycle of growth. Sekihara clearly identifies decay and its source: in the arrogance of the megacities and the actions of the caretaker state that cultivates generations of helpless, trapped, and dependent people. Kuni is the way forward precisely because it offers something for those who inhabit the forgotten rural communities as well as those who inhabit the megacities.

The two-loop theory urges us to draw lines between the collapsing system and the emerging alternatives. These lines matter greatly, even if the alternatives appear isolated and small. From Sekihara's and my point of view, the alternatives matter precisely because they are small. Using the pirate ships metaphor, the future depends on those on the sinking ship, who are invested in the old system, to glean new insights from those aboard the pirate ships, who are creating the new. These connections allow for those who are invested in the old system to decay with dignity. When the right time comes, they will integrate anything and everything from the new models. Adaptation may come too late, but it will ease the transition.

You might ask, do we care? If the old system is reckless and destructive in the case of food, biodiversity, human health, climate stability, and so on, should we not let it sink on its own, like the ships of the Spanish armada?

The problem is this: While in theory, pirates sail small ships and the empire sails big ones, most of us spend time on both vessels. It is difficult to live beyond the reaches of the empire. We participate in the decaying present and in the emerging future. I may strive to achieve food sovereignty, independent of large food distribution channels, to purchase provisions, but where will I find salt if the conventional distribution channels go dark? While we may fortify the alternative models so that they flourish, they are not yet strong enough.

For this reason, it is in our best interests to ease the transition. This is why the transfer of knowledge between the armada and the pirates matters. Yes, there are those defenders on the big ships who will do everything to undermine the credibility of the pirates. Also true, allies are aboard the armada. However, their loyalty to the existing system means that they are unable to recognize how that loyalty limits the bounds of thinkable thought. They cannot see that it is not enough to alter techniques. As pirates, we must learn to recognize these traits and not be distracted by them. Alas, it must be said, they possess skills that will be valuable when the time comes to abandon the old and embrace the new.

The longer my colleague continued to describe the two-loop theory to me, the more she brought me back to those meetings years earlier with farmers who had opted not to embrace the direct marketing model of farmers markets. Like those passengers aboard the Spanish armada who gaze from the deck of the ship out to the horizon, they catch glimpses of alternative models. They recognize the shapes and patterns of ships closer to them, whereas the ones far off may seem too distant and unfamiliar.

At the time, this new insight also confirmed for me an important position that strikes at the core of this book: There is a right size. While the declining system may scramble to persist, and in so doing will go out of its way to dismiss any and all alternatives, especially on the grounds that to follow these small efforts would be reckless and suicidal, these outbursts are scare tactics. For instance, industrial agriculture may remind us that they feed the world, and were we to abandon their model, there would be mass starvation. How many times have I encountered provocative questions like "What do you want, a farmers market on every corner?" Or "What do you plan to eat during the winter when your local food disappears?"

It is not our job to fix their problems. Nor is it our job to become like them, or even big like them. Rather, it is our job to make our models stronger, more lucid, more effective, and more attractive. When we succeed in becoming better versions of ourselves, then we make it easier for those who defend the realm to grow comfortable with the changes we offer.

While this may sound like boosterism from a coach to players at halftime to simply do better, it is actually a strategy. It is a call to resist the unhelpful yet nearly universal demand that every promising practice must scale up. If all we do is mimic the norms of the declining system, then we risk becoming like the old system. We will lose our tether and become lost at sea. Eagerly, invite crew from the figurative "ships of the armada" to board your pirate ship to share ideas; however, do not allow for these interactions to distract from the ultimate goal—to map out an alternative future.

Working within the context of what is broadly called the food movement, interventions vary greatly, from field to fork. Motivated as a consumer in the city, I saw the opportunity to rethink the point of sale by developing farmers markets—a strategy that leverages the point of sale as a teachable moment. The bet is that lessons learned at market work their way back to decisions made on the farm. Ideally, shoppers transfer knowledge about their preferences for flavor, crops, and varieties at the point of sale. Insights are returned to the farm to influence decisions about seeds, harvesting schedules, and so forth. Similarly, farmers inform consumers about seasonality and product availability. In both, the flows of information often deliver new information. For far too long, one has grown isolated from the other. As a point of leverage, it is not perfect. The intervention comes late in the process. A shopper cannot influence what you have already harvested. However, look forward to the next season for indicators that change has occurred. For this reason, others are drawn to other points of intervention: from land ownership and retention to postharvest handling and culinary knowledge in the kitchen. Each comes with its own strengths and weaknesses. Though they exist, it is uncommon for efforts to address all of the steps, from field to fork, holistically. Nearly every advocate (or pirate) has the appetite to act holistically, but few have the capacity or the resources to do so. In this respect, the two-loops model correctly describes emerging practices as isolated. Organizations like Slow Food provide a menu of intervention that is holistic. However, it is a heavy lift that is met with mixed results. More often than not, the needs outstrip the resources to succeed.

From community forestry to agroecology.

Mapping points of intervention is essential for measuring success and building a field. While Sekihara may not identify his work as strictly of the food movement, he understands the utility of food as a means to an end. Because he's a passionate woodworker, his first point of intervention was in the forest: to clean the debris from the forest floor and from the irrigation ditches. This led to efforts down the chain to processing and marketing, to reenergize employment with lumber. Disappointed with the results, Sekihara pivoted. No longer could he trust a marketplace that rewards larger scale and lower prices. He disengaged from a sectoral approach. No longer did he see his work as a timber strategy. Instead, timber is like food. It is an asset on which to trade. However, still smarting from his experience to produce lumber for a market that lacks an ethical compass, he bypasses the conventional industrial path and seeks out consumers who reward quality, authenticity, and ultimately relationships through experience.

Sekihara's forest-to-market journey brings conclusions about the kinds of markets to pursue in very close proximity to the ideas promoted by the field of agroecology. It advocates the fusing of traditional and scientific knowledge to integrate ecological thinking into agriculture. Among its high-profile advocates is Chilean-born agronomist Miguel Altieri. In 2016 I moderated a discussion with him at the Slow Food festival Terra Madre Salone del Gusto in Italy. Altieri insists that farmers must find alternative paths to bypass the conventional marketplace. He contends that, as Sekihara found with timber products, integrity gives way to meet price and demand. The norms and rules of the old system exert market pressures to compromise.

The agroecology critique that the market as it is cannot usher in the changes needed places it very much as a part of the emergence that the Berkana Institute's two-loops theory describes. Sekihara's Fan Club strategy—to forge commercial relationships that invite outsiders into the production process and to community rituals—clearly resides outside the existing distribution system. By design, it cannot be scaled up in the conventional marketplace.

While Sekihara may have begun with handcrafted wood products like bowls and tables fashioned from the lumber in the nearby forests, very soon the product line expanded to food—rice, pickled plums, and more. In order to successfully organize a coherent and professional agricultural enterprise, the community of producers themselves must be on board to share in the vision. A prerequisite for participating, they must be stable in order to function reliably for commerce. Based on Sekihara's description of the state of affairs in the villages, they were anything but stable. How did Sekihara stabilize the unstable? In order to get any of the work off the ground, the Fan Club balanced near-term community improvements (previously provided by local government) with long-term community goals. I really admire this approach. Perhaps because Sekihara lives in the community, he aggregates activities that span different sectors, and delivers near-term wins. He is able to bring neighbors along on a holistic process of reimagining lives and livelihoods. The niche efforts that people like me in food are able to accomplish are relegated to discussions about food, even if our desires are to transform society as a whole.

Places or people?

The taste of place is everything to the food movement. Granted, there may be no agreed-on definition of what that place is that we call local. Paradoxically, this may be the term's greatest strength. It allows everyone to be protagonists in imagining their locale. The locavore movement has come to define the brand of the food movement. From catchy concepts like the "hundred-mile diet" to the proliferation of local sourcing signs in supermarkets, "local" serves as a demarcation between local authenticity and placeless globalization. Or, to put it another way, it is the delineation between pirates and the armada. It is more than that. It also draws distinctions within the movement, between different ideological camps: the industrial agriculture reformers on the one hand, and the agroecology fundamentalists on the other.[1] While these may represent deep and irreconcilable divisions, the locavore imagination largely defines a goal that is cherished widely. It is akin to the emotional attachment to "organic" at the end of the twentieth century.

Local is a useful proxy for expressing opposition to and separation from the "global industrial agricultural complex." However, it is just that: a proxy. It does not communicate the entire story. Other useful indicators, like growing practices or fair labor policies, can easily be hidden. However, it does address an important indicator: the ecology of local economies. Money spent here stays here. Or, at least, it is more likely to remain here and generate local wealth.

Localization drives home more and more precise targets. Consider the rise in hyperlocal agriculture. The renaissance in backyard and community gardens, beekeeping, and urban fruit orchards expresses this drive. Also, consider the rise in urban commercial agriculture. Hothouse lettuce farms atop supermarket roofs, neighborhood community-supported agriculture enterprises, and even door-to-door egg delivery from urban chickens begin to measure food miles in feet and meters. These new social and commercial enterprises are cause for great celebration. With the introduction of each new moving part in a community's increasingly complex web of food options, a hopeful future appears ever more possible. Consumers begin to strike closer to Carlo Petrini's dream of coproducers, as they become more knowledgeable and active in their consumption habits. However, this drive to hyperlocalism also has the potential to cultivate an unfortunate race toward hyperpurism. One outcome the food movement does not need is to propel itself toward even greater enclaving.

Where will it end? Will we begin to debate the superiority of apples grown within the city limits versus ones just over the line? Already our urban versus rural tribalism is rotting what remains in civil peace and discourse. I contend that we need more bridging and less bonding. One of the food movement's secret weapons is its ability to trigger cognitive dissonance among participants. Sekihara engages urbanites from Tokyo who are drawn to his rural region, even if they are unable to express precisely why. They arrive seeking strong social ties. They may not be able to identify just how strong, but they take considerable pains to travel three or four hours by train. Meanwhile, farmers market vendors and shoppers meet on the tarmac (as it often is a parking lot or town square) of weak social ties. Their transactions are brief yet frequent. What I observed is this: Over a season, both sides undergo a cognitive dissonance. The rural farmers' expectations of the city people begin to clash with preconceived notions. The same occurs for the urban dwellers. Over repeat interactions, worldviews begin to shift. It is not easy to admit being party to stereotyping. As a result, little is made of these subtle shifts; however, in conversations with others involved with urban-rural linkages through food, I soon learned that my observation aligns closely with what other organizers observe, and importantly, begin to identify as a stated goal.

During the aftermath of Hurricane Katrina in coastal Louisiana, my organization led efforts to stabilize a community of commercial shrimping families who had previously directed some or all of their catch to alternative local markets. With the entire population dispersed all over the country, shrimpers needed new customers. Banded together beneath the banner of the White Boot Brigade (so named for the iconic white boots most fishers don), we developed retail and restaurant opportunities in New York City and San Francisco. During frantic sorties, we discovered that, when products and people are carefully vetted, place is no longer an essential proxy. The international fair-trade movement operates on this principle. However, it does not facilitate direct relations between supply and demand. As a purchaser of coffee, you will not meet your farmer. By contrast, we maintained direct contact between buyer and seller. All that we extended was the distance traveled.

As local food proponents, journalists asked us, had we learned anything? Indeed, we had. Between the vectors of people, place, and product, we discovered that our moral compass could accommodate relations without a shared affinity for place. Of course, solidarity should know no bounds. Prior to the devastating hurricane, at the farmers market, shoppers and vendors would search for commonalities. Maybe the shopper had fished somewhere near the shrimper's favorite haunt, or hunts on land near a farmer. However, the distance between Manhattan or San Francisco and coastal Louisiana provides few common threads. At that point, with the World Trade Center terror attacks only a few years prior, New York City chefs knew trauma. During meetings with chefs, many nodded in agreement when shrimpers described their fears for their families and their livelihoods. Aided by new communications technologies, sympathetic chefs and retailers, and generous transport partners, the White Boot Brigade broadened the reach of commerce and community for the shrimpers.

For Sekihara, the Fan Club draws wider community boundaries much in the way the White Boot Brigade does for the Louisiana shrimpers. He dismisses the narrowest boundaries of place: the villages too small to leverage resources and retain people. While the municipal authority of Joetsu, twenty minutes away, may hold actual jurisdictional power over the lives and livelihoods in the rural villages, the municipality acts more as a caretaker for communities that are entering into hospice. While residents of urban Joetsu may join in with the activities of the Kamiechigo Yamazato Fan Club, it is interestingly enough the residents of Sekihara's former town of Tokyo, four hours away, who imagine their future with the villagers.

The Fan Club administers a broader regional identity for those who live, farm, and make things in rural Joetsu. Their products are branded as originating from a place on the map that has no legal status. The origin is imagined, but the relationships are very real. It is one matter to invent a regional place, quite another to garner resources to develop it. More impressively, it is an entirely different matter to attract outsiders from the other side of the country to join in an experiment to restore dignity to a place where they will never live. More than an intentional community, kuni trades on place polygamy.

Yearning to Be Elsewhere

I remember playing in the yard in front of my childhood home in New Orleans during the summer of 1975. I was ten years old. I stood there watching a commercial airliner disappear into the clouds. It was like the one I had flown the summer before to stay with grandparents in England, where I always felt at home. I had this strange yearning to be there, not here. This is how I felt for most of my early life. Even today, I can replicate that strange trance, a deep longing to be there and not here.

So, when William Morrish, professor of urban ecologies at the New School for Social Research in New York City, introduced me to German sociologist Ulrich Beck's concept of place polygamy, my eyes lit up. I had not yet read the key passages in his 1997 book *What Is Globalization?* However, I knew immediately the deep pull of the concept—intellectually and emotionally.

Beck recognizes that globalization introduces so many new risks into our lives, personally and collectively. Interested in the unprecedented scale and ways that we shape and are being shaped by the ease that capital and people cross national borders, he refers to place polygamy as an outcome of what globalization has on our personal biographies. It is something that pushes us "toward globalization in personal life."[2] To illustrate this point, he describes a woman "who is eighty-four years of age, and who may therefore be called old":

> If the local register is to be believed, she has lived without a break for more than thirty years in Tutzing, on Lake Starnberg near Munich. A typical case of (geographical) immobility, one might say. But in fact our old lady flies at least three times a year to Kenya for several weeks or months at a time.... Where is she "at home?" In Tutzing? In Kenya? Yes and no. In Kenya she has more friends than in Tutzing and lives in a dense network of Africans and Germans, some of whom have their "residence" near Hamburg, while others "come" from Berlin. She also enjoys life more in Kenya, although she would not like to do without Tutzing altogether. In Africa she is invited to local people's houses and is generally well looked after. Her well-being in old age is based on the fact that she "is someone" in Kenya: she has a "family" there. In Tutzing, where is officially registered, she is a nobody. In her own words, she lives there "like a songbird."[3]

Granted, the old German woman possesses resources that allow her to travel. Hers is a luxury refugees cannot afford. Migrants travel for work and a new life because their old one has fallen apart. Beck is not suggesting that globalization is grand because it enables elderly people in the global North to afford meaningful lives during their declining years in places with weak economies. Rather, he is interested in how these changes shape our personal lives. They are proxies for how our relationships to places are changing.

In another illustration, Beck describes how the simultaneous marriage to several places creates a globalization of biography:

> Doris, who is forty years younger, is married to a (Muslim) Asian in Kenya, but she keeps going back to Germany to earn her money here or there (wherever one cares to see it), and to make sure all is well with the house and garden she owns in the Eifel mountains. She feels well in both places—which does not mean that the coming and going is not by now a little too much for her.[4]

Beck is writing at a time of triumphant and accelerated trade liberalization, technological innovation, and the dismantling of national labor markets. A loosening of national boundaries is disruptive. Poor people in the global South witness the arrival of new industries and livelihoods, just as they experience the dismantling of others. To meet these new opportunities, rural people clamor to urban areas. As these epicenters swell, workers find work. However, people can outnumber jobs, just as they can overwhelm social services and the safety net, if there is one.

As a result, people strike out for new horizons in search of opportunity. This occurs within and between national borders. One of the clearest indicators of place polygamy is the financial contribution migrants make to their home communities. These remittances not only play a major role in stabilizing the economies of home, but they also codify the dual commitments to place. At the time of publication, we are in the third decade of the twenty-first century. Few would argue that it is a good thing for desperate people to abandon home because it cannot sustain work with dignity. And yet, the flow of desperation continues. Families are dispersed across vast distances. So too are loyalties. Life becomes complicated. Where is home?

Prosperity is promised (or at least imagined). After all, this is what coaxed Tsuyoshi Sekihara from a dilapidated corner of rural Niigata to Tokyo's bright lights. Who can blame him? What would his prospects have been had he remained? Cities (and especially megacities) attract not only the desperate and the poor, but anyone who believes that unless they are part of the glamor and machinery of large places, they and their talents are doomed.

I think back to my return from studies in London to New Orleans. I felt like a loser, returning home. I discovered that very few of my classmates from high school were returning to a sinking city. Evidently, anyone who had a hint of a future fled for healthier economies. Only losers remain, while the winners fulfill the brain drain. I recall feeling embarrassed about my U-turn when I would encounter visiting classmates. It was even worse to enter into conversations with parents of classmates. They were itching to describe how smart their grown children were to find success anywhere but here.

These are hardly new dynamics. Young people have fled the constraints of home for success out there in the world long before the age of globalization. However, the political economy of globalization accelerates a tendency to flee to opportunities elsewhere.

Moreover, what is troubling is that globalization provides an intellectual basis for abandoning places and moribund economies. Mobility is deemed a strength: mobility in commerce and in people. In the United States, it is the acceptable norm that people, and especially young people, should follow prosperity. This kind of slash-and-burn economics leads to a development model that erects new towns, built to last long enough to handle the prosperity. As the horizon for prosperity grows nearer and nearer, the net result is a patchwork of half-built decaying towns that may house people but do little for civic life, democracy, and care for the land and water beneath our feet. The logical conclusion to this line of thought is troubling. If the assumption is that mobility is in my professional future, then it certainly reduces my need to become politically involved in my surroundings. By the time anything changes, I will be gone. While this may not be the singular cause for the degradation of democracy, no wonder so few citizens in the United States vote.

In this regard, kuni serves as a hopeful counterpoint to the nihilistic nomadic life I describe above. If the American norm reduces the desire or need to care for all places, the Fan Club's Rice Covenant offers the rituals and social contracts for individuals to truly care for and participate in more than one place.

Is Kuni Exceptional?

It is easy to become enamored with the coherent elegance of Sekihara's kuni. From the delicate Japanese joinery in the remodeled farmhouses' woodwork to the careful inclusion of outsiders for local Shinto festivals, I have difficulty believing that the North American food movement's track record in building community and bridges between communities matches up. While so much of the work I know or have developed myself seems so light and loose, Sekihara's is grounded. Perhaps this is the difference between starting points. When I began to reinvent public markets in New Orleans, I was able to build upon the fading memory of a market system; however, I had to start from scratch: new locations, new farmers, new brand identity, new organizational structure, new regulatory environment, and so forth. As a social entrepreneur, I assembled a team and identified locations in order to operate during discrete days and hours. The call-to-action amounts to this: Abandon your old habits and join something new. Fear not, you maintain your independence and identity, and can choose to return or not. Through community organizing, we had established some fragile and new relationships with potential participants. However, we were on the outside of lives and livelihoods, seeking ways in. Once inside, with degrees of trust established, then we build something new together.

By contrast, Sekihara had taken the leap of faith and already committed his future to be entwined with his neighbors by taking that J-turn and landing in rural Joetsu. He could have remained passive and remained on his own property, but he did not. Through his practical volunteer actions, he demonstrates his desire to take on forest work that benefits no one in particular but everyone in general. In this regard, Sekihara is already on the inside. He organizes where he lives and asks his neighbors to share their strong social ties with outsiders. From his own account, these agreements with neighbors did not just happen. He had to demonstrate that by coming together, their risks are shared. By sharing these risks with selected outsiders, their risks are shared even more widely.

In the United States, and maybe also in Japan, Sekihara would face pressures we know in food and community development to scale up. In order to reach wider audiences and have greater impact, abandon the strong social ties. They require too many resources to sustain.

This is not to suggest that Sekihara's Regional Management Organization, the Kamiechigo Yamazato Fan Club, is unlike any effort in the United States. In fact, it is my hope that this book will tease many like-minded experiments from out of the woodwork to proclaim that they have been doing the same work for years. If so, this would only further illustrate my point that the community development work that gets through the white noise and receives public recognition tends to be larger and urban, relies on weak social ties, and resembles patterns that look more like the old system we want to replace than the new ones we imagine replacing the old.

It is the imagination in Sekihara's work that brings him closer to us. How he threads together so many different activities and tendencies into a holistic reinvention of place may indeed be unique: voluntourism, the twelve functions of a Regional Management Organization, leadership development, and the matching of youth with the elderly. His mind is expansive, but his actions are deliberate. This is why his voice is so welcome in the field of rural community development. His ability to assemble many different activities as one is unusual. However, his use of imagination is on display the world over.

In farmers markets large and small, our movement envisions our rules and regulations more as manifestos for regional cooperation than as procedures manuals. Or consider the tiny Welsh town of Hay-on-Wye. Its population swells from two thousand inhabitants to eighty thousand for ten days each spring for its annual literary festival. It took the imagination of a few individuals to rewrite the narrative of an otherwise unremarkable yet picturesque town on the Welsh-English border. This raises a question: Does a festival remake a town? I can only imagine that the imposition of tens of thousands of visitors during a compressed period of time each year is both welcome and unwelcome news for locals. Regardless, the Hay Festival of Literature and Arts does point to the imagination of a few individuals (in this instance, members of one family) to rewrite the narrative of place and why it matters on the map.

Similarly, the case of Fogo Island off the coast of Canada's Newfoundland describes what happens when an inhabitant returns many years later (in a U-turn) to the fishing village with resources accrued from a successful professional career in industry on the mainland. Zita Cobb returned to an otherwise forgotten and shrinking community, much like others in the region that once rode the waves of the cod fisheries industry. She returned a success, and she ploughed her success into redeveloping the island and its economy around the arts, a beautifully designed hotel, and retreat cabins. Each is adorned with motifs designed by and emulating local skilled crafts and traditions. While few can replicate the wealth and control that Cobb's Shorefast Foundation has upon the development of the Fogo Island Inn, her forceful reimagination of place on a theme that welcomes outsiders is a close match to Sekihara's kuni. Cobb's project celebrates and employs the artistry of locals who remain, just as it does for those who, like Cobb herself, had to leave in order to pursue a life and livelihood. Fogo Island has become something of a catchphrase for a design-specific type of rural reinvention. This both intimidates those without the kind of philanthropic leadership Cobb is able to provide just as it inspires. Sekihara may not possess the kind of financial resources that floats Cobb's Fogo Island work; however, he possesses remarkable skills to leverage resources through charisma and hard work. These are attributes that continue to show themselves all over the world, especially when a place that is supposed to be doomed to history reemerges as a bright spot, a hopeful counterweight to the steamroller of homogenization, scale, and speed that blankets the planet.[5]

Perhaps it is the situation in sunny yet depressed southern Italy that captures the political nature of Sekihara's work the most. It is a region that triggers many starry-eyed conversations among North Americans for its provocative campaign to sell houses that cost one euro. How desperate are we to change our lives?

Relocation and New Forms of Insurance

Are we at a turning point? How many urban dwellers are "this close" to chucking it in for somewhere and something else? According to the *Asahi Shimbun*, "one in four teleworkers mull ditching Japan's big cities for rural areas."[6] Many ponder whether the pandemic ushers in lasting changes to how work and offices are structured. Will the temporary ability to work remotely remain? Or, for that matter, will work itself remain? Few people dispute the sanity of the twentieth-century norm to commute vast distances to work in order to pay for the domicile that is underutilized during the day. Factor in additional pressures and the precarious balancing act that keeps people in place might simply give way. Questions that have been rumbling beneath the surface for years (How are we to manage rising rent, health care and day care costs, and an obscene competition for seats in good schools?) will move above ground, perhaps resulting in action. If we are indeed on the cusp of major demographic changes, the business model for urban neighborhoods, with the plethora of gyms, restaurants, cafés, and boutiques, will not hold.

This is not to suggest that the pandemic that began in 2020 will necessarily yield a wholesale back-to-the-land movement. After all, the very reasons why Italian towns like Cinquefrondi are driven to boast provocative real estate deals hinge upon chronic desperation. Maybe you have been lured into purchasing one-euro homes in breathtakingly beautiful yet decaying agricultural villages in Italy. An internet sensation, this strategy has kept many dreamers awake late at night researching maps of forgotten corners of Italy in hopes that some sort of future there is possible. I know that I am one of those place polygamists. Small rural communities have been left behind and gutted of core assets, hospitals, schools, and a sense of why they exist in the first place. Is there enough there, there?

Are towns and villages ready to receive newcomers? Do they have the infrastructure to meet newcomers' expectations? Or worse, if they bend over backward to do so, will they sacrifice locals' needs for newcomer demands? If they're not ready, how can Sekihara's RMO model inform the need for inclusive yet independent leadership in these fragile places? If, on the other hand, yes, they are prepared, then how willing are newcomers to conform to the prevailing norms of these small places?

In Cinquefrondi mayor Michele Conia's description of Operation Beauty (the program to redevelop the bedraggled parts of town), he describes his as "a land of cross-contamination and cross-civilizations." He also uses language like "melting pot" and a "welcoming people."[7] These indications are hopeful pronouncements, but are they just slogans? Closer inspection indicates that there is much more to this welcoming posture.

Like Sekihara's rural edge of Joetsu, Cinquefrondi presents itself as open to newcomers. Conia has brought his municipality of 6,450[8] into the Network of Solidarity-based Communities (RECOSOL). Other small Calabrian towns—short on skills, youth, and a tax base—provide asylum seekers with housing and job training. This strategy is intended to prevent municipal death.[9] RECOSOL has been championed by the nearby Calabrian municipality of Riace and has earned justifiable international attention for its approach to forging social cohesion. This is a far cry from the ugly and inhospitable signals sent to desperate refugees by Italy's xenophobic political party the League (Lega in Italian).

These admirable municipal strategies, pursued by Calabrian towns, echo Sekihara's kuni. Desperate to pump new blood into fading rural regions, they open doors to outsiders. However, the value proposition is more than just cheap land deals. In Japan, take the Kamiechigo Yamazato Fan Club's Rice Covenant. It sets in motion a pathway for you to become more and more invested in a regional community's future. Purchase a kilogram of rice directly from the community of farmers. Do it again. In short, begin as a shopper, return as a friend. Through each transaction, you learn that they are trading more than just rice. They are trading membership in a fading community and a way of life that is heroically reviving and reinventing itself. This kind of transaction is therefore meant to be transformative for both parties. Importantly, it is a process, not a quick transaction. As a shopper, you purchase rice and earn a spot on a database that gives you access to village life and activities supported by the RMO.

Access means you get first dibs to purchase the highest quality of the season's hand-harvested rice or umeboshi plums. This is where the relationship begins and, perhaps, remains for some time. However, through repeat visits, you become more familiar to the region's residents, and they to you. Join the harvest and

Shinto religious festivals. Attend a forestry workshop; visit the cultural museum for a demonstration; stay in the inn. In short, become a part of the social fabric of the community.

Leaders like Sekihara and Conia should know one another. Both articulate a desire to forge deeper relationships between their communities and the outside world. On one hand, Cinquefrondi turns to those even more desperate than the town itself—refugees via RECOSOL. On the other, it turns to those seeking a good real estate deal, via Operation Beauty. Referencing the latter, Mayor Conia boasts how his town has more than good and decaying building stock to offer. As in rural Joetsu, Cinquefrondi stages agricultural fairs and religious festivals. One is Spinati, which features the barefooted "thorn men."[10]

With both, municipalities target audiences seeking security. In Italy, refugees seek a long-overdue safe haven from war. In Japan, there are too few refugees to compare. However, with regard to the urban dwellers who keep returning to these far-flung places, they too seek security. Be it to live permanently or visit frequently, Sekihara's Rice Covenant articulates it best when it assures those whose lives feel increasingly tenuous in the "new wildness" of the megacity that safety can be found in kuni.

Are We Ready to Take the Leap?

It is safe to assume that not everyone is ready to take the leap. The Rice Covenant recognizes that the leap requires an extended process of assimilation to grow into new lives with new norms. This is the insurance policy outsiders pay into during their numerous visits and via their online purchase of products. When conditions become unbearable, you are welcome to find safe haven. Maybe it will be an earthquake, a pandemic, or the next round of layoffs to trigger the moment you abandon the culture of gigantism and embrace the right-sized community envisaged by Tsuyoshi Sekihara.

With civil discourse literally tearing societies apart, we have choices. Run to the barricades and take issues head on: gun ownership, reproductive rights, animal rights and human rights, decolonization and racial equity. Locked into fixed positions that define the contours of cultural wars, we lack proximity in each other's lives. Not only do we not know one another, we have too few opportunities to address the distance and knowledge. No wonder a bumper sticker is little more than the opening salvos for an argument. Instead of rushing toward conflict, what if we were to seek commonalities in order to carve out the social space for exploration? While Sekihara's quiet experiment in rebuilding community in forgotten rice farming communities in Japan may seem worlds away from the familiar and fading small communities in North America, kuni's value is in its inspiration. The Kamiechigo Yamazato Fan Club makes sense in Japan because it draws from patterns familiar to the Japanese. Sekihara may urge organizers elsewhere to follow many of the tactile lessons he learned. He may even insist on considering the twelve functions of holistic community organizing via Regional Management Organizations. However, most of all, kuni is a gift to the wider world because its political project refuses to mimic the dangerous and destructive norms that contribute to community decay. You can run to the drama of conflict, or you can quietly build the bridges on which relationships are forged.

From global pandemic to climate crisis, from globalization to the numbing and cumulative effects of fractured urban lives, the call to small and fragile places is both a call to wildness and an invitation to community.

Notes

Chapter 1

1 Leopold Kohr, *The Breakdown of Nations* (1957; repr., Cambridge, UK: Green Books, 2001), 87.

2 This is the slogan of iconic New Orleans snowball stand Hansen's Sno-Bliz. The message is inscribed on the ice-shaving machine that the business's founder, Ernest Hansen, built in the 1930s.

3 Benedict Anderson, *Imagined Communities: Reflections on the Origin and Spread of Nationalism* (London: Verso, 1983), 15. Drawing heavily on the insights of historian Hugh Seton-Watson in his book *Nations and States: An Enquiry into the Origins of Nations and the Politics of Nationalism* (Boulder, CO: Westview Press, 1977).

4 Bretton Woods, New Hampshire, was the site for the agreements made by the winning Allies at the close of World War II as to how to construct international relations once peace was reached. It is a framework for the conduct of international monetary, financial, and trade relations. These regimes have shaped life since World War II: the United Nations, the International Monetary Fund, the World Bank, and others.

5 Mark Jefferson, "The Law of the Primate City," *Geographical Review* 29:2 (April 1939), 226–32. doi:10.2307/209944.

6 From an informal interview with Tozaburo Sato, April 14, 2019, in Yamagata, Japan, organized by the Japan Society and the Japan NPO Center. It is worth noting that while Sato recognized the then-Imperial Japanese government's support for rural agriculture, his remarks were in no way an indication of a longing for an imperial past. Rather, that was the last time he felt that rural Japan caught the eye of central government.

7 From an informal interview with Michio Kimura, April 14, 2019, in Yamagata, Japan, organized by the Japan Society and the Japan NPO Center.

8 BBC Future, "Dunbar's Number: Why We Can Only Maintain 150 Relationships," October 1, 2010, BBC.com. https://tinyurl.com/bcjfpnbx. See Robin Dunbar, *How Many Friends Does One Person Need?: Dunbar's Number and Other Evolutionary Quirks* (London: Faber & Faber, 2010).

9 A very large city, usually meaning with at least ten million inhabitants. With this class of city accelerating in number, there is a growing concern that they concentrate wealth, people, resources, and our ability to imagine what normal life is.

10 United States Department of Agriculture (USDA) Economic Research Service, *Local Food Systems: Concepts, Impacts, and Issues*, ERR 97, May 2010. https://tinyurl.com/yckmwvpk.

Chapter 5

1 Masanobu Fukuoka, *The One Straw Revolution: An Introduction to Natural Farming* (New York: New York Review of Books, 1978), 115.

2 Fukuoka, *The One Straw Revolution*, 1.

3 "Brides for Bumpkins: Japan's State-Owned Version of Tinder," *The Economist,* October 3, 2019. https://tinyurl.com/324wapd9.

4 Wynn Rosser, "A Call for Philanthropy to Invest in Rural America," FSG.org, October 10, 2018. https://tinyurl.com/3269nha6.

5 Rosser, "A Call for Philanthropy to Invest in Rural America."

6 Personal interview with Ben Burkett in Petal, Miss., May 9, 2019.

7 United States Department of Agriculture (USDA) Economic Research Service, "Farming and Farm Income," January 6, 2022. https://tinyurl.com/yv5y68xa.

8 USDA Economic Research Service, "Farming and Farm Income."

9 USDA Economic Research Service, "Farming and Farm Income."

10 Danial Martinus, "65 Percent of the World Drives on the Right. But Why Do Some Countries Stick to the Left?," Mashable SE Asia, August 28, 2020. https://tinyurl.com/2p8yvnla.

11 It is common to take a train to remote wilderness with the expressed purpose to immerse yourself in nature and soak up the smells and sounds of the flora and fauna free from disruptive technology and the chattering masses. This is what is called forest bathing. While many of us may take a hike in the wilderness, with clear beginnings, ends, and marked trails, forest bathing is itself the end. While in the United States the pursuit to enjoy nature has long excluded people of color, the joy of walking or "rambling" in Britain is historically and specifically associated with equity: the right to ramble on private property. The Japanese love for forest bathing shares this quest for access with British rambling; however, it also removes the necessity of having

geographical goals in mind. Instead, you head to the forest to leave civilization and reconnect with nature.

12 Tokyo hosted the Olympics in 1964, an event that brought major architectural changes to Tokyo and the beginning of an opening up to the world again after the postwar rebuilding. The proposed, and then COVID-canceled, 2020 Tokyo Olympics was expected by government and industry to play a similar role after the 2011 Great Tohoku Earthquake and tsunami.

13 Alex Kerr, *Lost Japan: Last Glimpse of Beautiful Japan* (Melbourne, Australia: Lonely Planet, 1996), 26.

14 Alana Semuels, "Can Anything Stop Rural Decline?" *The Atlantic,* August 23, 2017. https://tinyurl.com/m2vfzy5z.

15 *Voluntourism* is hybrid term that gained considerable traction in the wake of Hurricane Katrina in New Orleans in 2005, when vast numbers of tourists began to travel to the recovery zone in order to volunteer. These acts of generosity have existed for years; however, this particular disaster seemed to strike a national mood—hungry for meaningful endeavors, seeking places that seem genuine and authentic, if not tragic—just as it coincided with the rise of the experiential economy, in which short-term rentals, Airbnb, and everything from sofa-surfing, gap years, and the increased interest in WWOOFers began to dominate the public imagination. *WWOOFers* refers to the volunteer farm labor organized by the Worldwide Opportunities on Organic Farms.

Chapter 10

1 I cannot help but be reminded of the divisions with the German Green Party, between the Realos (realists) and the Fundis (fundamentalists).

2 Ulrich Beck, *What Is Globalization?* (Malden, MA: Blackwell, 2000), 72.

3 Beck, *What Is Globalization*, 72–73.

4 Beck, *What Is Globalization*, 73.

5 Writings about Fogo Island are not difficult to find; however, a great place to start is the documentary film *Strange & Familiar: Architecture on Fogo Island* (Marcia Connolly and Katherine Knight, directors, video, First Run Features, 2015).

6 Tomohiro Yamamoto, "1 in 4 Teleworkers Mulling Ditching Japan's Big Cities for Rural Areas," *Asahi Shimbun*, June 22, 2020. www.asahi.com/ajw/articles/13479412.

7 Silvia Marchetti, "Cinquefrondi: The 'COVID-free' Italian Town Selling $1 Houses," CNN Travel, June 10, 2020. https://tinyurl.com/2p9e5ksk.

8 "Cinquefrondi, in Reggio di Calabria (Calabria)," City Population, May 15, 2021, https://tinyurl.com/2p8j547j.

9 Donatella Loprieno, Anna Elia, and Claudio Di Maio, "Integration Governance in Italy: Accommodation, Regeneration and Exclusion," GLIMER Project, University of Calabria, 2020. https://tinyurl.com/2p9dydtc.

10 Marchetti, "Cinquefrondi."

Index

About the Authors

Photo credit: Fumiko Miyamoto

Tsuyoshi Sekihara: Founder and former executive director, Kamiechigo Yamazato Fan Club, Joetsu, Niigata.

Sekihara grew up in rural Niigata Prefecture and left to pursue a career in Tokyo as a commercial facilities designer. In 1996 he returned to Niigata and soon became the director of a woodwork cooperative, where he continues to serve as a consultant. In 2000 the organization received the Ministry of Agriculture, Forestry and Fisheries Award; however, he came to recognize that the market for quality wood products is vulnerable to global pricing pressures. This led him to seek more holistic strategies, most notably the Kamiechigo Yamazato Fan Club, which he founded with the expressed goal to revive the hilly and mountainous areas in the greater Joetsu area. Under Sekihara's leadership, Kamiechigo, a comprehensive Regional

Management Organization (RMO), became a model for the Ministry of Agriculture, Forestry and Fisheries' program that provides local governments with financial support to hire local support staff. The role of the eight full-time and thirty part-time staff is to monitor the well-being of residents. The RMO manages an annual operating budget of $500,000 and serves twenty-five communities. Sekihara is also a regional empowerment creation adviser with the Ministry of Internal Affairs and Communications, a member of the RMO Study Group at the Ministry of Internal Affairs and Communications, and a trustee of the Kodo Cultural Foundation. While he has written extensively, very little is published beyond practitioners' circles. He devotes much of his off-work hours to designing furniture, knives, and renovating Edo Period farmhouses. In October 2019 he traveled to New York City to feature in three days of symposia at the Japan Society—Embrace Rural with Janet Topolsky of the Aspen Institute; Professor William Morrish, director of the New School for Social Research's Urban Ecology Program; and Kathleen Finlay of Glynwood Center for Regional Food and Farming.

Richard McCarthy: Founder and former executive director, Market Umbrella, New Orleans, and former executive director, Slow Food USA.

McCarthy grew up in New Orleans. After studying political science at the American University in London (Richmond) and the London School of Economics, he returned to put ideas into action. In 1995 he developed the practitioners' think tank Market Umbrella to cultivate public markets for public good. He developed the Crescent City Farmers Market as the organization's primary base of operations. Innovations include a workers' cooperative with public housing residents; the nation's first farmers market health incentive pilots, which led to the development of the USDA GusNIP program; an international research fellowship to measure the human, social, and financial capital in markets; and many critical recovery efforts after Hurricane Katrina. In all his work, two important themes prevail: to facilitate trusting relations between urban and rural communities, and to instigate lasting social change by creating opportunities for behavior change to come first. Ideas follow. McCarthy's work became increasingly global through partnerships with the UN World Summit on Sustainable Development, Project for Public Spaces, and Slow Food. In 2013 he left New Orleans to direct Slow Food USA for six years. Today, he works with Meatless Monday, Slow Food International, FAO, the World Farmers Market Coalition, and others to cultivate community through food. He addresses audiences all over the planet, has published chapters in two books, and has produced short films and radio programs. He currently sails under the name Think Like Pirates.

About North Atlantic Books

North Atlantic Books (NAB) is a 501(c)(3) nonprofit publisher committed to a bold exploration of the relationships between mind, body, spirit, culture, and nature. Founded in 1974, NAB aims to nurture a holistic view of the arts, sciences, humanities, and healing. To make a donation or to learn more about our books, authors, events, and newsletter, please visit www.northatlanticbooks.com.